MEETING THE NEEDS OF PEOPLE WITH DISABILITIES THROUGH FEDERAL TECHNOLOGY TRANSFER

HEARING

BEFORE THE

COMMITTEE ON SCIENCE
SUBCOMMITTEE ON TECHNOLOGY
U.S. HOUSE OF REPRESENTATIVES

ONE HUNDRED FIFTH CONGRESS

FIRST SESSION

JULY 15, 1997

[No. 26]

Printed for the use of the Committee on Science

U.S. GOVERNMENT PRINTING OFFICE

44–186CC WASHINGTON : 1997

For sale by the U.S. Government Printing Office
Superintendent of Documents, Congressional Sales Office, Washington, DC 20402
ISBN 0-16-056046-2

COMMITTEE ON SCIENCE

F. JAMES SENSENBRENNER, JR., Wisconsin, *Chairman*

SHERWOOD L. BOEHLERT, New York
HARRIS W. FAWELL, Illinois
CONSTANCE A. MORELLA, Maryland
CURT WELDON, Pennsylvania
DANA ROHRABACHER, California
STEVEN SCHIFF, New Mexico
JOE BARTON, Texas
KEN CALVERT, California
ROSCOE G. BARTLETT, Maryland
VERNON J. EHLERS, Michigan**
DAVE WELDON, Florida
MATT SALMON, Arizona
THOMAS M. DAVIS, Virginia
GIL GUTKNECHT, Minnesota
MARK FOLEY, Florida
THOMAS W. EWING, Illinois
CHARLES W. "CHIP" PICKERING,
 Mississippi
CHRIS CANNON, Utah
KEVIN BRADY, Texas
MERRILL COOK, Utah
PHIL ENGLISH, Pennsylvania
GEORGE R. NETHERCUTT, JR.,
 Washington
TOM A. COBURN, Oklahoma
PETE SESSIONS, Texas

GEORGE E. BROWN, Jr., California RMM*
RALPH M. HALL, Texas
BART GORDON, Tennessee
JAMES A. TRAFICANT, Jr., Ohio
TIM ROEMER, Indiana
ROBERT E. "BUD" CRAMER, Jr., Alabama
JAMES A. BARCIA, Michigan
PAUL McHALE, Pennsylvania
EDDIE BERNICE JOHNSON, Texas
ALCEE L. HASTINGS, Florida
LYNN N. RIVERS, Michigan
ZOE LOFGREN, California
LLOYD DOGGETT, Texas
MICHAEL F. DOYLE, Pennsylvania
SHEILA JACKSON LEE, Texas
BILL LUTHER, Minnesota
WALTER H. CAPPS, California
DEBBIE STABENOW, Michigan
BOB ETHERIDGE, North Carolina
NICK LAMPSON, Texas
DARLENE HOOLEY, Oregon

TODD R. SCHULTZ, *Chief of Staff*
BARRY C. BERINGER, *Chief Counsel*
PATRICIA S. SCHWARTZ, *Chief Clerk / Administrator*
VIVIAN A. TESSIERI, *Legislative Clerk*
ROBERT E. PALMER, *Democratic Staff Director*

SUBCOMMITTEE ON TECHNOLOGY

CONSTANCE A. MORELLA, Maryland, *Chairwoman*

CURT WELDON, Pennsylvania
ROSCOE G. BARTLETT, Maryland
VERNON J. EHLERS, Michigan
THOMAS M. DAVIS, Virginia
GIL GUTKNECHT, Minnesota
THOMAS W. EWING, Illinois
CHRIS CANNON, Utah
KEVIN BRADY, Texas
MERRILL COOK, Utah

BART GORDON, Tennessee
EDDIE BERNICE JOHNSON, Texas
LYNN N. RIVERS, Michigan
DEBBIE STABENOW, Michigan
JAMES A. BARCIA, Michigan
PAUL McHALE, Pennsylvania
MICHAEL F. DOYLE, Pennsylvania
ELLEN O. TAUSCHER, California

---

*Ranking Minority Member
**Vice Chairman

CONTENTS

MEETING THE NEEDS OF PEOPLE WITH DISABILITIES THROUGH FEDERAL TECHNOLOGY TRANSFER

TUESDAY, JULY 15, 1997

U.S. HOUSE OF REPRESENTATIVES,
COMMITTEE ON SCIENCE,
SUBCOMMITTEE ON TECHNOLOGY,
Washington, DC.

The Subcommittee met at 2:05 p.m., in room 2318 of the Rayburn House Office Building, Hon. Constance A. Morella, Chairwoman of the Subcommittee, presiding.

Mrs. MORELLA. I'm going to call to order the Technology Subcommittee of the Science Committee.

I want to welcome you all to this afternoon's hearing on the federal technology transfer applications of assistive technologies.

Assistive technologies are being used to increase, maintain, and improve the functional capabilities of individuals with disabilities. For the 49 million people in the United States who have disabilities, as well as the Americans who are able-bodied, our Nation's federal research and development investment, especially from our federal laboratories around the country, has yielded a tremendous number of quality of life enhancements.

These include some of the most state-of-the-art advances in assistive technologies, some of which will be demonstrated at today's hearing.

Each day, research and development programs and our Nation's over 700 U.S. federal laboratories produce new knowledge, processes, and products.

Often technologies and techniques generated in these federal laboratories have commercial applications if further developed by the industrial community. As a result, federal laboratories are working with the U.S. business industry and state and local governments to help them apply these new capabilities to their own particular needs.

Through this technology transfer process, our federal laboratories are sharing the benefits of our national investment in scientific progress with all segments of our society.

By spinning off and commercializing federally developed technology, the results of our federal research and development enterprise are being used today to enhance our Nation's ability to compete in the global marketplace.

It seems clear that the economic advances of the 21st Century will be rooted in the research and development performed in our

Nation's laboratories and is becoming more dependent upon the continuous transfer of technology into commercial goods and services.

Today, we will focus on one such success story of our federal technology transfer efforts, assistive technologies.

The majority of the approximately 1,000 companies in the assistive technologies field are small-to-medium-size companies which could benefit from the same access, partnership, and collaboration of activities with the federal laboratory system that's become far more commonplace in fields such as automotive manufacturing, aeronautics, energy and environmental research.

I look forward to hearing more about assistive technologies from our distinguished panel today. And I want to welcome our witnesses to the Committee.

In conjunction with this afternoon's hearing, there's been an all-day exhibition of the latest developments in assistive technologies held in room 2168. I highly recommend participating in the exhibit demonstrations for those of you who may not have done so before they close at 4 p.m.

I have already been there and thoroughly enjoyed participating in the various exhibitions. And I want to commend the companies and our federal laboratories, and those who have commercialized, for the work that they've done.

I also want to recognize Tyrone Taylor and Greg Fine of the Federal Laboratory Consortium for all of their hard work in assisting the Committee with the exhibition and the hearing.

I now will turn to my Ranking Member, the distinguished gentleman from Tennessee, Mr. Gordon, for any opening remarks he may have.

Mr. GORDON. Thank you, Chairwoman Morella.

Good afternoon. I want to welcome everybody to this hearing.

First, I want to congratulate the Federal Laboratory Consortium (FLC) on their continued efforts to transfer technology from the federal labs to the private sector. I'm a strong supporter that any federal investment in research and technology development should be exploited to the fullest extent possible.

Often technologies developed for specific lab missions can be adapted and utilized in the private sector to a much wider benefit than the lab's original intent. The FLC provides the bridge which allows these successes.

Today's hearings highlight this concept. Private companies have adapted government developed technologies to assist people with disabilities. As a result, taxpayers reap many times over their original investment.

I'm well aware of the importance of the need for increased attention to development of assistive technologies. My alma mater, Middle Tennessee State University, has one of the largest populations of students with disabilities in the southeast United States. This fall, more than 650 students with disabilities will be enrolled at MTSU.

MTSU is consistently striving to improve the educational experience of these students and utilizing assistive technologies is a major component of their program.

I also would like to add that in addition to the witnesses we'll hear today, Oak Ridge National Lab has also been active in the area of assistive technologies. This year, Oak Ridge engineers were Discovery Award finalists for their work on technology that has been used to improve power wheelchairs.

I want to thank the witnesses for appearing before the Subcommittee and I look forward to hearing their comments and demonstrations of assistive technologies.

Mrs. MORELLA. I'd like to now recognize the gentleman from Michigan, if he has any opening comments.

Mr. Ehlers?

Mr. EHLERS. Thank you, Madam Chairwoman. I do not.

Mrs. MORELLA. Thank you.

Since it's the policy of the Science Committee to swear in all its witnesses, will the members of the panel please stand, raise your right hand and respond.

[Witnesses sworn.]

Mrs. MORELLA. The report will designate that the answers are in the affirmative.

Thank you.

We have six witnesses who are going to appear here in the one panel this afternoon.

Our first witness is Dr. Katherine Seelman, Director of the National Institute of Rehabilitation and Research, Department of Education.

Dr. Seelman was named by President Clinton as the Director of NIDRR and has been an administrator, researcher, and advocate on disability policy and was kind enough to take me through the exhibition today.

Next to Dr. Seelman is Mr. C. Dan Brand, current Chair of the Federal Laboratory Consortium for Technology Transfer. Mr. Brand is also the Associate Director for Technology Advancement at the National Center for Toxicological Research in Arkansas.

As a matter of fact, he used to be my constituent. Then when I was sworn into Congress, he moved to Arkansas. I don't know why.

Bruce Webbon will also testify. He represents the NASA Ames Research Center in California, where he is the Chief of the Commercial Technology Office.

Mr. Webbon will be able to tell us of the assistive technology advances resulting from technology originally developed at Ames.

The final three witnesses on our panel are from industry and have been active in assistive technologies in that field. These industry representatives come from large, medium, and small-sized companies, and will give us their first hand accounts of their work with assistive technologies.

Mr. Steve Jacbos comes from Dayton, Ohio, the home of NCR's Corporate headquarters, where he is an executive office technology consultant.

Mr. David Hershberger is Vice President of Product Development for Prentke Romich Company in Wooster, Ohio.

Mr. Joseph Lahoud is the founder and President of LC Technologies based in Fairfax, Virginia.

Welcome everyone to today's hearing.

Just an aside. Copies of the testimony will be made available in the form of braille or an ASCII format on diskette following the hearing by FLC and NIDRR. There is a sign up sheet outside the hearing room.

So now we'll commence with the testimony starting with Dr. Seelman.

Just remember that anything you submit in the way of testimony will in its entirety be part of the record, so feel free to synopsize your comments.

Dr. Seelman?

TESTIMONY OF KATHERINE SEELMAN, DIRECTOR, NATIONAL INSTITUTE OF REHABILITATION AND RESEARCH, DEPARTMENT OF EDUCATION, WASHINGTON, DC

Ms. SEELMAN. My name is Katherine D. Seelman and I bring you greetings from Secretary of Education, Richard Riley, and the Assistant Secretary of the Office of Special Education and Rehabilitative Services, Judy Heumann.

I am Director of the National Institute on Disability and Rehabilitation Research, and chair the Interagency Committee on Disability Research.

I would like to express my appreciation for your leadership in technology transfer and for inviting me to testify. I am testifying on the prospects for potential collaboration between NIDRR, the disabled community, and the federal laboratories in the areas of assistive technology.

The importance of assistive technology to disabled people is summarized in the following quotation:

"For Americans without disabilities, technology makes things easier. For Americans with disabilities, technology makes things possible."

Activities many of us take for granted, such as moving from place to place, may be impossible for someone with paralysis. But a wheelchair makes the impossible become possible.

The well-known symbol of disability seen everywhere clearly reflects this concept. A person using assistive technology now has the possibility of moving about independently.

I would now like to give you a brief overview of NIDRR. NIDRR's purpose is to support rehabilitation research to improve the lives of individuals with disabilities.

My written testimony contains greater detail, but we are the lead federal agency in research, development, and deployment of assistive technology.

We support a program of rehabilitation research at the level of about $70 million, and we also support a $36 million effort in assistive technology authorized by the Technology Related Assistance Act.

I am a user of assistive technology in the form of hearing aids and am also speaking on behalf of the estimated 16.5 million other Americans who use or who would benefit from assistive technology.

Some of NIDRR's past accomplishments in assistive technologies through its rehabilitation engineering research centers include working with Microsoft to include accessibility features in Windows

95, and contributing substantially to launching the now thriving augmentative communication device industry.

What assistive technology is in use today?

In addition to the devices I mentioned earlier, there are relatively simple devices such as canes and walking sticks, as well as more technologically sophisticated devices such as digitally programmable hearing aids, voice input and output computers, artificial limbs made of space age materials and perhaps someday a visual prosthesis.

There is a broad spectrum of assistive technology. At one end of the continuum are devices used by relatively small numbers of individuals with disabilities. These are sometimes called orphan devices.

At the opposite end of the spectrum are devices used by a large number of people, both disabled and non-disabled that are universal in nature.

Orphan technology or orphan devices are typically designed to restore physical function or retrofit a commercially available piece of equipment so that someone with a disability can use it.

An example of this is a doorbell activated fan to alert a person who is both deaf and blind that someone is at the door.

Universally designed devices for the mass market have accessibility features built in at the front end so that virtually anyone can use them.

A well-known example of this would be the inclusion of close captioned decoder in all television sets. This is useful not only for deaf and hard-of-hearing persons, but all persons in noisy areas such as airports and restaurants.

There are a number of trends that I've noted in my written testimony, but they are trends of particular importance to the disability community. They include the rapid and pervasive change in mainstream technology. They can create not only new opportunities but new barriers.

For example, people who are blind working with computers may find themselves no longer able to perform their jobs if they cannot use computers based on graphical icons.

Another major concern is aging. Older Americans can use assistive technology to prolong the time they can live independently in the community and to reduce the cost of long term care.

What is the future of assistive technology?

This is what brings us together today: the ability of the federal laboratories to play a significant role in future assistive technology development.

Most technology transfer models suggest the need for an intermediary of some sort. NIDRR is ideally placed to serve in such a role.

Engineers at NIDRR's rehab engineering centers can speak both the language of their fellow engineers in the federal laboratories and the language of the disabled community.

NIDRR has recognized exchange of information between NIDRR constituents and the federal laboratories as perhaps being the most important area to address.

NIDRR has funded the Consumer Assistive Technology Transfer Network to help in this regard.

Further cooperation and partnership between NIDRR, the disability community, and the federal laboratories might include continuing work on information exchange, providing additional resources in the form of engineering staff time to seek out potential technologies for transfer, conducting pilot projects or reciprocal exchanges of staff engineers involving individuals with disabilities and individuals from minority backgrounds in the process.

Congresswoman Morella and members of the Subcommittee, you have an impressive record of fostering tech transfer. I hope with your leadership, new ways can be found to expand our ability to work collaboratively to help the federal laboratories transfer their technology in that very important area of assistive technology to the benefit of people with disabilities.

Let me close by returning to the beginning quote. "For Americans without disabilities, technology makes things easier; for Americans with disabilities, technology makes things possible."

For an individual with a disability, assistive technology can make possible what would otherwise be impossible. Turning impossibilities into possibilities is perhaps one of the finest outcomes of technology transfer.

Thank you.

[The prepared statement of Ms. Seelman follows:]

Written Testimony
of

Katherine D. Seelman, Ph.D.
Director
National Institute on Disability and Rehabilitation Research
Office of Special Education and Rehabilitative Services
U. S. DEPARTMENT OF EDUCATION

on

Meeting the Needs of People with Disabilities
Through Federal Technology Transfer

to

United States House of Representatives

Committee on Science
Subcommittee on Technology
Tuesday, July 15, 1997
2:00 p.m.
2318 Rayburn House Office Building
(Exhibits 2168 Rayburn)

INTRODUCTION:

Chairwoman Morella and members of the Subcommittee on Technology, my name is Katherine D. Seelman. I bring you greetings from the Secretary of Education, Richard Riley, and the Assistant Secretary of the Office of Special Education and Rehabilitative Services, Judy Heumann. I am the director of the National Institute on Disability and Rehabilitation Research (NIDRR), and chair of the Interagency Committee on Disability Research (ICDR). I would like to express my appreciation for your leadership in technology transfer, and for inviting me to testify before you today at this hearing on Meeting the Needs of People with Disabilities Through Federal Technology Transfer. I am testifying on the prospects for potential collaboration between NIDRR and the Federal Laboratories in the area of assistive technology.

The importance of assistive technology to disabled people is summarized in the following quotation:

> "For Americans without disabilities, technology makes things *easier*. For Americans with disabilities, technology makes things *possible*."

(From <u>Study on the Financing of Assistive Technology Devices and Services for Individuals with Disabilities</u>, National Council on Disability, March 1993)

Activities of life that many of us take for granted, such as moving from place to place, may be impossible for someone with paralysis. However, the addition of assistive technology, in this case a wheelchair, makes the impossible become possible. The widely recognized symbol of disability, that can be seen everywhere, clearly reflects this concept: a person using a wheelchair (assistive technology) now has the possibility of moving from place to place independently.

NIDRR OVERVIEW:

NIDRR is located in the Office of Special Education and Rehabilitative Services (OSERS) in the Department of Education. The purpose of NIDRR is to support rehabilitation research and the use of such research to improve the lives of individuals with physical and mental disabilities, especially those with severe disabilities. We are the lead federal agency in research, development, and deployment of assistive technology. We support a comprehensive program of rehabilitation research at a level of about $70,000,000 per year. NIDRR's research activities most relevant to the hearing today are the sixteen Rehabilitation Engineering Research Centers (RERCs). We also support related work through Small Business Innovative Research (SBIR) contracts and through discrete research grant projects. The Program Directory for the research supported by NIDRR is available on the World Wide Web at

http://www.naric.com/naric/nidrr96/intro.html

NIDRR administers a $36,000,000 effort in assistive technology deployment authorized by The Technology-Related Assistance for Individuals with Disabilities Act (PL 100-407). There are Tech Act Programs in all 50 states.

Another one of my responsibilities as Director of NIDRR is chairing the ICDR, which is charged with promoting coordination and cooperation among Federal departments and agencies conducting rehabilitation research. The ICDR's Technology Subcommittee is actively working to promote and coordinate these activities.

NIDRR's community of interest includes individuals with disabilities and their families, researchers, rehabilitation service providers, and manufacturers and distributors of assistive technology across the nation.

One of NIDRR' s main areas of focus is research in the application of engineering and technology to assure equality of opportunity, full participation, independent living, and economic self-sufficiency for individuals with disabilities. Some examples of NIDRR's past accomplishments through its Rehabilitation Engineering Research Centers include working with Microsoft to include accessibility features in Windows '95, and contributing substantially to launching the now thriving augmentative communication device industry. Mr. Hersberger, of the Prentke Romich Company, will also testify today and tell you more about augmentative communication devices, which allow individuals who cannot speak to communicate vocally.

ASSISTIVE TECHNOLOGY TODAY:

What assistive technology is in use today? A definition of assistive technology is attached, but I would like to give a few specific examples. In addition to the wheelchair and augmentative communication devices mentioned earlier, assistive technology includes relatively simple devices, such as canes and walking sticks, as well as more technologically sophisticated devices such as digitally programmable hearing aids, voice input/output computers, artificial legs and hands made of space-age materials, and perhaps someday, a visual prosthesis. I hope you saw some of the exhibits prior to the hearing, or will visit them afterward.

There is a broad spectrum of assistive technology. At one end of the continuum are devices used by relatively small numbers of individuals with disabilities. At the opposite end of the spectrum are devices used by large numbers of people, both disabled and non-disabled.

Those pieces of assistive technology that are specialized for use by individuals with a particular disability are sometimes called "orphan devices". Orphan devices are disability specific devices that are used only by a small number of individuals with disabilities. These devices are designed with the concept of restoring a physical function, or retrofitting a commercially available piece of equipment so that someone with a disability can use it. An example of an orphan device is a device used by an individual who is both deaf and blind, in which a doorbell activates one or more fans to alert the individual when a visitor is at the door.

Currently, an estimated 15.6 million people in the US either use some type of specialized assistive technology or have reported they would benefit if they did use assistive technology (LaPlante et al, <u>Technology and Disability</u>, vol 6, pp. 17-28, 1997).

At the opposite end of the spectrum are devices that have been developed through the process of "Universal Design". In this process, devices for the mass market have accessibility features built in at the front end, so that virtually anyone can use them. A well-known example of this would be the inclusion of the closed-caption decoder in all television sets, making receipt of captioned broadcasting available to millions for only pennies per set. Captioned broadcasts are useful not only for deaf and hard-of-hearing persons, but for persons learning to read English and all persons in noisy areas such as airports and restaurants. Another example is the information kiosk that provides the same information in both visual and auditory formats. Universal technology is also represented by ramps that can be used by both individuals in wheelchairs and persons pushing baby strollers, and by lever style handles that make opening doors and turning faucets easier for all individuals. The universal approach typifies the concept of "fixing" the environment. Mr. Jacobs will be discussing Universal Design in his testimony today.

It is more difficult to obtain solid estimates of the market sizes and user populations of universally designed technology.

MARKET AND DEMOGRAPHIC TRENDS:

Looking to the future, population trends as well as trends in technological developments will require attention. Massive and pervasive changes in mainstream technology will create not only new opportunities, but also new barriers. For example, blind people who work with computers may find themselves no longer able to perform their jobs if they cannot use new computer systems that are based heavily on graphical icons.

With respect to trends in the globalization of technology, and the concomitant rapid economic changes, many experts believe that the U.S. and international markets for assistive technology will:
>> experience significant growth in the years ahead,
>> be very receptive to the introduction of new devices,
>> need to respond to consumer demand for improvements to the quality of
    existing assistive technology devices, and
>> experience increasing competition.

At present, there are approximately 2,500 companies, the vast majority of them small, that produce assistive technology in the U.S. They are listed in ABLEDATA, a national assistive technology database developed with NIDRR support, available on the World Wide Web at:

http://www.abledata.com/index.htm

These companies constitute the core of an infrastructure to further the transfer of technological discoveries to assistive technology.

Demographic studies reveal the trend to the aging of the U.S. population, and that is a trend experienced by other industrialized nations. An older population has greater assistive technology needs than a younger population. Older Americans use assistive technology to prolong the time they can live independently in the community and to reduce the costs of long-term care.

PLANS FOR THE FUTURE:

Where will our country's future assistive technology come from? How will the United States respond to these changes and trends? This is what brings us to the main point of the hearings today -- the ability of the Federal Laboratories to play a significant role in assistive technology development for the future. Together, NIDRR and the Federal Laboratories are looking for ways to bridge the gap between the high technologies within the Federal Laboratories, the opportunities for assistive technology development, and the needs of people with disabilities.

Based on past interactions with Federal Laboratories, researchers in NIDRR's Rehabilitation Engineering Research Centers have identified the need for better access to, and exchange of, information and expertise. While great strides have been made in cataloging the technological resources of the Federal Laboratories, the problem of describing vast amounts of extremely complex technology is, in itself, a daunting challenge.

We have experienced advances in our exchange of information between NIDRR constituents and the Federal Laboratories.   For example, NIDRR has funded the Consumer Assistive Technology Transfer Network (CATN). The CATN is charged with increasing consumer involvement in assistive technology transfer by linking the fifty-six State programs funded under the Technology-Related Assistance for Individuals with Disabilities Act with the sixteen Rehabilitation Engineering Research Centers (RERCs), and the seven hundred Federal Laboratories nationwide. Interacting with CATN permits engineers, developers, and researchers to review new product inventions and problems submitted by consumers, and permits consumers to review the potential value of new technologies produced by engineers, developers, and researchers. More information about CATN can be obtained at

>   http://www.rt66.com/catn.org

As a result of various CATN activities, the Federal Laboratory Consortium has funded assistive technology focus groups at NIDRR's Technology Transfer RERC. The purpose of these groups is to look at the possible relevance to users of various advances in assistive technology.

A model for future successful cooperation and partnership between NIDRR projects and the Federal Laboratories includes:

>> continuing to work on information exchange,
>> providing additional resources, in the form of engineeringstaff time, to seek out
   potential technology for transfer,
>> conducting pilot projects or reciprocal exchanges of staffengineers and scientists
   between NIDRR's RERCs and the Federal Laboratories, and
>> involving individuals with disabilities and individuals from minority backgrounds in
   technology transfer.

CONCLUSIONS:

Congresswoman Morella, the Subcommittee on Technology of the House Committee on Science has an impressive record of fostering technology transfer. I hope that with your leadership, new ways can be found to expand our ability to work collaboratively to help the Federal Laboratories transfer their technology into the very important area of assistive technology, to the benefit of people with disabilities. Let me close by returning to a quote I used in the introduction:

> "For Americans without disabilities, technology makes things *easier*. For Americans with disabilities, technology makes things *possible*."

For an individual with a disability, assistive technology can make <u>possible</u> what would otherwise have been <u>impossible</u>. Turning <u>impossibilities</u> into <u>possibilities</u> is perhaps one of the finest outcomes of technology transfer.

ATTACHMENT

<u>Definition of assistive technology</u>
The term assistive technology device means any item, piece of equipment, or product system, whether acquired commercially, off- the-shelf, modified or customized, that is used to increase, maintain, or improve functional capabilities of individuals with disabilities (Excerpt PL 100-407) .

<u>Common Usage</u>
Technologies used in education, rehabilitation, and independent living to help, change, or train are now commonly grouped into the generic term "assistive technology." People of all ages with physical, cognitive and communication disorders, or a combination of disabilities may benefit from the application of assistive technologies.

EXAMPLES

<u>Interaction with subject matter and instructional materials</u>
Computers with adaptive switches and keyboards that substitute for normal keyboard use 'or conventional handwriting; audio tape players, braille displays or print magnifiers for students who are blind or visually impaired.
<u>Communication</u>
Persons with speech and/or hearing disabilities are able to transmit and/or receive communication. Communication boards, speech synthesizers, modified typewriters, head pointers, text to voice software, voice to text software, and telecommunication devices for the deaf, text telephones, CCTV's.
<u>Mobility and active movement</u>
The use of electric or conventional wheelchairs for full body movement; modifications of vans for travel; crutches, canes and walkers for support and stability; or canes used by pedestrians who are blind or visually impaired.
<u>Control of equipment</u>
Switches for people who have limited control over voluntary movements can be activated by touch, sound, voice, light pointers, and movement of the body for computers, television, and home appliances. Self-help skills Devices that assist in daily living and independence skills: modified eating utensils, adapted books, pencil holders, page turners, dressing aids, and adapted personal hygiene aids.
<u>Body support, alignment, and positioning</u>
Adapted seating, standing tables, seat belts, braces, cushions and wedges to maintain posture, and devices for trunk alignment that assist people in maintaining body alignment and control so they can perform a range of daily tasks.
<u>Modification of the work environment</u>
Modifications of desks and work tables to accommodate wheelchairs, computer modifications for alternate input systems, "talking" instrument displays for the blind or visually impaired, and automatic door openers.
<u>Leisure time, recreation activities, and appreciation of the arts.</u>
Devices include guide rails in bowling alleys for people who are blind, special prostheses that assist persons with amputations to participate in sports, computer decelerators that slow down arcade type games, and audio description for movies, sporting, and cultural events.

UNITED STATES DEPARTMENT OF EDUCATION

OFFICE OF SPECIAL EDUCATION AND REHABILITATIVE SERVICES
NATIONAL INSTITUTE ON DISABILITY AND REHABILITATION RESEARCH

August 4, 1997

Michael Bell
Subcommittee on Technology, Committee on Science
U.S. House of Representatives
Suite 2320 Rayburn House Office Building
Washington, DC 20515-6301

Dear Mr. Bell:

In response to Congresswoman Morella's letter of 17 July 1997, enclosed please find grammatical and transcription corrections to my oral testimony for the July 15th, 1997 hearing on "Meeting the Needs of People With Disabilities Through Federal Technology Transfer."

I would also like to make the following statement in this letter to be appended to the official record.

Dr. Bruce Webbon, one of the witnesses from NASA, has correctly pointed out in his oral testimony that: "The resources are often available so that new money and new resources do not necessarily need to be provided, but direction does have to be provided because these efforts, in my experience, are in addition to your primary job. So that needs to come from the very top to allow you to do it."

I concur with Dr. Webbon's assessment that efforts in technology transfer from the Federal Laboratories into the field of assistive technology are going to be most successful if undertaken by individual researchers (or perhaps small teams of researchers) as a sideline to their primary job. The key here is getting the word out that the higher levels of management within the Federal Laboratories are in favor of this occurring, and that individual researchers can expect help and encouragement when they make such attempts.

Certainly these hearings went a long way in publicizing that the Subcommittee on Technology looks favorably on these efforts. I hope this view will be adopted over time within the Federal Laboratories themselves.

Sincerely yours,

Katherine D. Seelman, Ph.D.
Director
enclosure: corrections
cc: The Honorable Constance A. Morella

Our mission is to ensure equal access to education and to promote educational excellence throughout the Nation.

KATHERINE D. SEELMAN is an administrator, researcher and advocate in disability policy. President Clinton named Dr. Seelman as Director of the National Institute on Disability and Rehabilitation Research (NIDRR). NIDRR is a Federal agency that oversees a $70 million discretionary research budget, as well as program that fosters the use of assistive technology in the homes and workplaces of persons with disabilities in all fifty states and U.S. territories. As part of her responsibility, Dr. Seelman is also chairperson of the Interagency Committee on Disability Research, which has been established to promote coordination and cooperation among Federal agencies conducting rehabilitation research programs.

Interested early in telecommunications policy and accessibility issues for persons with disabilities, Dr. Seelman has advanced scientific endeavor in these areas. In a similar manner, she brought to the NIDRR an awareness of the emergency of non-traditional disability populations, reinvigorating the agency's research agenda and targeting it toward the perceived needs of the 21st century. She has promoted the visibility of disability studies and Participatory Action Research constructs designed to advance the interests of persons with disabilities throughout academic study and research.

During her tenure as Director of NIDRR, Dr. Seelman has served as chairperson of the Research Committee for the 1996 Paralympic Games in Atlanta. Internationally, she has spearheaded a major initiative on educational technology, as part of the U.S.-Japan Common Agenda. This year she represented the Administration in meetings in Japan with high level officials in the areas of Special Education and Rehabilitation. She traveled to India and Israel on similar representational visits. Recently, she addressed the World Congress of Rehabilitation International in New Zealand.

Dr. Seelman began her career as a community organizer and teacher. Prior to her appointment to NIDRR, Dr. Seelman, who is hard of hearing, served in leadership positions in such well-known institutions as the National Council on Disability, the Administration on Developmental Disabilities, the Massachusetts Commission for the Deaf and Hard of Hearing, the Gallaudet Research Institute and the Office of Technology Assessment of the U.S. Congress.

Dr. Seelman earned her doctorate in public policy from New York University in 1982. During her career, she has received numerous awards, including a National Science Foundation Assistantship and a distinguished Switzer Fellowship. She has been a Switzer Scholar with the National Rehabilitation Association and she was selected to deliver the prestigious Switzer and Schaeffer Lectures. In 1995 she was honored by her alma mater with induction into the Hunter College Alumni Hall of Fame. She is listed in Who's Who in America and Who's Who of American Women.

Mrs. MORELLA. Dr. Seelman, that's a great quote and your testimony, including the written testimony that we have, really confirms that.

We've been joined by Ms. Tauscher from California and Ms. Rivers from Michigan.

Now we'll turn to Mr. Brand.

TESTIMONY OF C. DAN BRAND, CHAIRMAN, FEDERAL LABORATORY CONSORTIUM FOR TECHNOLOGY TRANSFER, ASSOCIATE DIRECTOR FOR TECHNOLOGY ADVANCEMENT, FOOD AND DRUG ADMINISTRATION, NATIONAL CENTER FOR TOXICOLOGICAL RESEARCH, JEFFERSON, AK

Mr. BRAND. I'd also like to acknowledge and thank both the Chair and the members of this Subcommittee for your invitation to speak on this very important subject, and discuss how our federal laboratories can help meet the needs of people with disabilities.

I'd also like to commend you, Congresswoman, and your members of the Subcommittee for your knowledge of the federal laboratories in your respective districts back home.

I would like to give you a little brief overview of the consortium and how we fit into the assistive technology network.

The consortium consists of 711 member research and development laboratories and centers from our 17 federal agencies and departments, certainly a Nationwide network of laboratories that provide the forum for those laboratories to develop some strategies and opportunities for moving the government technologies into the marketplace.

We also bring to the laboratories the potential users of government-developed technologies both in the private sector, as well as state and local governments, a very big piece of this.

We also develop and test transfer methods, address some barriers to the process, and provide training and highlight grassroots transfer efforts, and emphasize national initiatives where technology transfer has a role.

Labs represent a vast reservoir, as you know, of technology and expertise. This reservoir, we believe, can be tapped to aid the disabled as well as the able bodied communities in our country.

The FLC has recently developed a new strategic plan for the 21st Century, to reach our vision of maximizing the return on the R&D from this Nation's investment.

In implementing this plan, we've identified six focus areas that are designed to do so. Two of these, I think, are most appropriate for the assistive technology arena optimizing the diverse resources that are available in the laboratories and creating innovative partnerships.

We believe that the FLC can assist in transferring the technologies out of the laboratories to the assistive technology community.

The FLC is committed to enhancing the partnership opportunities between the laboratories and the private sector which includes the assistive technology community.

We believe that these increased partnership opportunities will lead to a more efficient leveraging of our federal resources that are available both in and out of the federal R&D system.

While individual laboratories from the departments, many of them represented here today, have worked with the assistive technology community over the years, our overall organization's commitment actually began with our participation in last year's 1996 International Para-Olympics and Abilities Expos.

The para-olympic activities include many things, among them sponsoring round table discussions involving people with disabilities and certainly the assistive technology manufacturing community.

Since last year, we have actively promoted and have been engaged in a number of activities supporting the Department of Education's NIDRR laboratory programs.

These activities conducted include an unprecedented technical session on assistive technology at our last two National Technology Transfer meetings.

Partnering with NIDRR on the development of the Consumers Assistive Technology Network and developing a multi-agency presentation for the NASA/FLC Technology 2007 Conference to be held in September in Boston this year.

Madam Chairwoman, I think it's important to emphasize what you said last year to our laboratories when you provided a vision of what our role might be and I'd like to quote from your statement to our laboratories last year.

The federal labs must refocus their missions on long-term, often high risk R&D frequently through facilities which are beyond the financial reach of industry and academia, and through the application of multi-disciplinary teams of scientists and engineers, the appropriate lab role with industry is one in which a cooperative R&D agreement is signed and implemented so that the labs are accomplishing one of their four primary missions; national security, energy, environmental and basic sciences, while seeking industrial participation only where it is needed to accomplish the primary mission.

Within that framework that you charged the federal laboratories with, I believe that the laboratories can assist in moving the technologies into the assistive technology community.

The majority of these companies are small to medium size firms which could benefit from the same access of partnerships and collaborations with the laboratory systems that have become commonplace, as you mentioned earlier, in the automotive manufacturing and biotechnology, aeronautics and energy and environmental research arenas.

Many of the basic technology needs of the assistive technology community are in areas such as devices for people with physical, sensory, and expressive, communication, and cognitive disabilities.

These technology focus areas represent many of the same research areas consistent with our R&D mission of the laboratories related to the technologies for space exploration, national security and our defense.

The recent changes that we've had in the technology transfer legislation which were introduced by you, Madam Chairwoman, serve to further promote, facilitate and give incentive to the public and private sector to develop these strategic partnerships, and to accomplish the missions of these laboratories.

It's through such partnerships that industry is able to identify broader application of federal laboratory technologies for the commercial market, including our assistive technology business community.

It is the goal of our organization to promote and facilitate an increase in these partnerships involving the assistive technology community through the use of the mechanisms that currently exist and that you have supported: Personnel exchanges, cooperative research and development agreements, use of facilities and laboratory expertise already available and existing in the federal system.

We need to ensure that these small business have efficient, reliable access to the system where it helps to fulfill the mission of those laboratories, while at the same time providing the Nation the opportunity for an increased return on our R&D investment.

The FLC and the NIDRR each have a unique mandate to serve the public. This synergy of mission needs to be developed further.

One of the six focus areas, optimizing our diverse resources, I spoke about a little earlier, has heightened relevance, I think, in this situation.

It's in the best interest of our Nation to ensure that we leverage to the fullest extent possible, our federal investment, where appropriate, to improve the quality of life for persons with disabilities.

In this environment of scarce resources, it's also critical that each dollar is stretched as far as possible. There needs to be a push to evaluate and assess technologies' universal design potential for an assistive technology application.

The federal R&D system needs to commit, at the very least, to explore areas of cooperation with this assistive technology community.

We have begun several new actions to further this endeavor, but there are four things that I think the consortium can begin immediately to start the process of assisting the NIDRR in its mission:

One is that the organization continue to promote the awareness and benefits of the assistive technologies universal design through the federal laboratory system.

The consortium, in conjunction with NIDRR, can begin to identify the technical needs of the NIDRR laboratories and the manufacturing community, and identify those member laboratories that can help.

Third, we can exploit and adapt the use of the mechanisms that the Congress and the Administration have provided through key legislation such as the 1995 Technology Transfer and Advancement Act, Small Business Innovative Research and Manufacturing Extension Programs, and the Assistive Technology Community Programs.

With your support, Madam Chairwoman, and that of the Subcommittee members, we can convey to the laboratory leadership, the importance of considering the incorporation of assistive technology design and development issues as it relates to the mission of their respective laboratories.

It is our organization's hope that these near-term actions will lead to a meaningful, long-term relationship and economic results that will ultimately benefit not only those people with disabilities, but society as a whole.

Finally, I just want to thank you for this opportunity to speak here today on behalf of the federal laboratories, and commend you, Madam Chairwoman, and members of the Subcommittee for your interest and leadership in this increasingly important subject of assistive technologies, and especially in the technology transfer arena from our laboratories.

Thank you very much.

[The prepared statement of Mr. Brand follows:]

Federal Laboratory Consortium

"Meeting the Needs of People with Disabilities Through Federal Technology Transfer"

C. Dan Brand, FLC Chair

Prepared for the Committee on Science
Subcommittee on Technology
U.S. House Representatives

July 15, 1997

**"Meeting the Needs of People with Disabilities
Through Federal Technology Transfer"**

**C. Dan Brand, Chair
Federal Laboratory Consortium**

Prepared for the Committee on Science
Subcommittee on Technology
U.S. House of Representatives
The Honorable, Constance Morella, Chair
July 15, 1997

Introduction
It is a privilege to appear before you today to discuss how the
Federal Laboratories can help meet the needs of people with
disabilities. I wish to acknowledge and thank both the Chair and
members of the Subcommittee for your invitation to speak on this
very important subject.   Before I discuss how we feel the FLC can
aide in the transfer of Federally funded technologies to the Assistive
Technologies (AT) community, I would like to provide a definition of
AT. Assistive Technologies are defined as any device, item, piece of
equipment, or product system, whether acquired commercially off the
shelf, modified or customized, that is used to increase, maintain or
improve functional capabilities of individuals with disabilities. I would
now like to give you a brief overview of the Federal Laboratory
Consortium for Technology Transfer (FLC).

FLC Overview
The FLC consists of over 700 member research/development (R&D)
laboratories and centers from seventeen federal agencies and
departments. The FLC is the nationwide network of Federal
Laboratories that provides the forum to develop strategies and
opportunities for moving Government technologies to the
marketplace. The FLC brings its member laboratories together with
potential users of government-developed technologies in the private
sector as well as state and local governments. We also assist in the

interagency transfer of technologies. We develop and test transfer methods, address barriers to the process, provide training, highlight grass-root transfer efforts and emphasize national initiatives where technology transfer has a role. The Federal Laboratories represent a vast reservoir of technology and expertise, a reservoir we believe can be tapped to aide the disabled and able bodied communities.

The FLC has been in existence in some form since 1974. The passage of the Stevenson-Wydler Technology Innovation Act of 1980 provided increased stimuli for the growth and development of the FLC. The Federal Technology Transfer Act of 1986 provided a Congressional Charter and funding mechanism creating the FLC as we know it today. The National Technology Transfer and Advancement Act of 1995, introduced by the Chair and co-sponsored by Representative George Brown, provided a number of significant enhancements for industry and the Federal Laboratories. This legislation also reinforced the importance and appropriate role of technology transfer programs and activities within the Federal Laboratory System.

The FLC has recently developed a new strategic plan for the 21st Century to meet our member's future needs. In implementing this Plan we have identified six strategic focus areas designed to help the FLC achieve our vision. A copy of our Strategic Plan (figure 1) and the six strategic focus areas (figure 2) is included for your review. Two of the six areas I believe, are directly related to how the FLC can assist in transferring technology to the AT Community. These two areas are, "Optimizing Diverse Resources" and "Creating Innovative Partnerships."

The FLC is committed to enhancing partnership opportunities between the Federal Laboratories and the private sector, which includes the AT Community. We believe that these increased partnership opportunities will lead to a more efficient leveraging of the resources that are available, both in and out of the Federal R&D System. While individual laboratories have worked with the AT Community over the years, our organizational commitment began with our participation in the 1996 Atlanta International Paralympics and Abilities Expo. Our Paralympics activities included for example, sponsoring roundtable discussions involving people with disabilities and the manufacturing community. Since August of last year, we have actively promoted and been engaged in a number of activities supporting the Department of Education's National Institute on Disability and Rehabilitation Research (NIDRR). These activities included conducting unprecedented technical sessions on AT at our last two National Technology Transfer Meetings, partnering with NIDRR on the development of the Consumer's Assistive Technology Network (CATN), and developing a multi-agency presentation for the Technology 2007 Show to be held September 22 -24, 1997 in Boston. Attached for your review is a summary of some of these activities (figure 3).

Madam Chair, last year in a speech to our FLC membership you provided a vision of what our role might be: I quote:

"...Federal labs must refocus their missions on long-term, often high-risk R&D, frequently through facilities which are beyond the financial reach of industry and academia, and through the application of multi-disciplinary teams of scientists and engineers... The appropriate lab role, with industry, is one in which a cooperative R&D agreement is designed and implemented so that the labs are accomplishing one of their four primary missions (national security, energy, environmental and basic sciences), while seeking industrial participation only where it is needed to accomplish the primary mission."

Within this framework, I believe the FLC member labortories can further assist in moving government technologies into the AT community, thus ultimately benefiting the citizens of the United States.

Based upon a recent report "Chartbook on Disability In the United States, 1996" , it becomes quite apparent that as we age, the level of severe disabilities increases from a low of 3.3 % of the population age 15-24 to 41.5% of the population age 75-84. I would suspect that this trend will continue as the nation continues to age.

Page 3
C. Dan Brand
AT Testimony
House Tech. Cmte.
July 15, 1997

AT Issues Background
As the nation continues to define the role and determine priorities of federally funded Science and Technology (S&T) activities, we believe it is imperative that Assistive Technologies be part of that debate. There are numerous planned and serendipitous scientific and engineering discoveries resulting from Federal R&D investment that have yielded quality of life enhancements for Americans who are able bodied as well as disabled.

There are currently 49 million people in the United States who have disabilities. Five percent of those people who could benefit from the use of an assistive technology use them. Although this is currently a $26.5 billion industry and growing, due to the aging process, it is an area that has largely not benefited from technology transfer activities within the Federal Laboratory System. Current projections indicate that this market of products and services will continue to grow at a rate of 10-12% annually for the foreseeable future.

Presently there are approximately 2,500 companies in the AT field, which is dominated by three major competitors capturing slightly less than 10% of the $26.5 billion market. The majority of companies are small to medium size firms which could benefit from the same access, partnership and collaboration of activities with the Federal Laboratory System that has become far more commonplace in fields such as automotive manufacturing, aeronautics, and energy and environmental research.

Many of the basic technology needs of the AT community are in areas such as devices for people with physical sensory and expressive communication and cognitive disabilities. These technology focus areas represent many of the same research areas consistent with the R&D mission of our laboratories related to technologies for space exploration, national security and defense.

Past Roadblocks to Success

A solid history of synergism does not exist between the FLC member laboratories and the NIDRR laboratories. NIDRR's sixteen, small, primarily University based, labs have often felt ignored and marginalized by their larger Federal counterparts. Efforts on the part of the NIDRR labs to work with other members of the Federal system have seen little success. A significant cause of this has been a general lack of understanding on the part of the major labs as to the opportunities that exist for collaboration between themselves, NIDRR and the AT manufacturing sector and the value added benefits from such collaborations. We believe that the FLC can play a major role in increasing this awareness.

Recent Changes in Legislation
The recent changes to the Technology Transfer legislation, which were introduced by the Chair, serve to further promote, facilitate and incentivize the public and private sector to develop strategic partnerships to accomplish the mission of the agencies. It is through such partnerships that industry is able to identify broader applications of Federal Laboratory technologies for the commercial market. We believe this approach is equally viable for the AT business community. Therefore, it is the goal of the FLC to promote and facilitate an increase in partnerships involving the AT Community through the use of the mechanisms which currently exist. i.e., personnel exchanges, Cooperative Research and Development Agreements (CRADAs), use of facilities and laboratory expertise. We need to ensure that small businesses have efficient, reliable access to the Federal R&D System where it helps to fulfill the mission of the laboratory, while providing the nation the opportunity for an increased return on its R&D investment.

Some Recent Success Stories
There have been numerous successful technology transfer efforts in the AT area. Examples include:

<u>Department of Defense</u>
- Use of Global Positioning Satellite System technology by Arkenstone in Sunnyvale, California to assist individuals who are

visually impaired.

Department of Transportation

- Talking Signs - A system which uses infrared transmitters and receivers to provide orientation to low vision or learning disabled individuals.

Veterans Administration (VA)

- VA/Seattle below the knee prosthesis that has been designed and fabricated using advanced computer aided design and manufacturing technology through a collaborative effort with Aulie Devices, Inc. of Redmond, Oregon.

National Aeronautics and Space Administration (NASA)

- A software program developed for research to evaluate the physiological and behavioral effects of flight systems on pilots has been adapted to become part of a system that will help educators communicate with severely disabled students. NASA's Langley Research Center has teamed with the private sector and the Montgomery County (Ohio) Board of Mental Retardation and Developmental Disabilities to find a way to objectively measure what information their disabled students are gaining from their environment so an education curriculum could be designed for them.

FLC/NIDRR Synergy
The FLC and NIDRR each have a unique mandate to serve the public. NIDRR is specifically charged with developing Assistive Technologies, and it is clearly within our mandate to assist NIDRR. This synergy of mission needs to be developed and enhanced. One of the six focus areas I spoke about earlier has heightened relevance here. It is in the best interest of our nation to ensure that we leverage to fullest extent possible the Federal investment, where appropriate, to improve the quality of life for persons with disabilities. This is important for a number of reasons.

First, is the simple fact that demographics alone indicate that the number of people requiring some form of AT is only going to increase. As the percentage of the U.S. population over 65 increases in the future, there will most certainly be an increase in the call for AT devices. It is paramount that R&D be done early to reduce the lag time between supply and demand.

Second, this increase in need will come at a time of decreasing resources, both in dollars and people. In this environment of scarce resources, it is critical that each dollar be stretched as far as possible. There needs to be a push to evaluate and assess a technology's "universal design" potential for an AT application. The Federal R&D System needs to commit, at the very least, to explore areas of cooperation with the AT Community.

This leads to the other area I mentioned earlier, to "Create Innovative Partnerships." The federal labs need to look outside the box during the assessment of technologies. Old ideas and practices need to be reviewed to ensure that access to technologies is not being limited by past ideologies and ideas. AT research and development needs the kind of attention that Biotech R&D enjoys today. The Federal Research System needs to understand and value the profound impact to the quality of life of the American public that Assistive Technologies can have to the disabled and able bodied community.

FLC Actions
There are four things that the FLC can begin immediately to start the process of assisting NIDRR in its mission.

1. The FLC will continue to promote the awareness and benefits of Assistive Technology universal design throughout the Federal Laboratory System.

2. The FLC, in conjunction with NIDRR, can begin to identify the technical needs of the NIDRR laboratories and the AT manufacturing community and then identify FLC member laboratory contacts who can help. It has been said that Technology Transfer is a "contact sport," putting individuals

together to talk and share ideas should begin to enhance the relationship between the FLC membership, NIDRR and the business community.

3. Exploit and adapt the use of mechanisms the Congress and Administration have provided through key legislation such as the National Technology Transfer and Advancement Act of 1995, Small Business Innovative Research and Manufacturing Extension Programs to the Assistive Technology community.

4. With your support madam Chair and that of the Subcommittee members, we can convey to the laboratory leadership the importance of considering the incorporation of Assistive Technology design and development issues as it relates to the mission of its respective laboratories.

It is our hope that these near-term actions will lead to meaningful, long-term relationships and economic results that will ultimately benefit not only those people with disabilities, but society as a whole. It is important to remember that as we age, or in a split second, due to an unforeseen catastrophic accident, anyone sitting in this room could require some form of Assistive Technology and if you needed it, wouldn't you want it as soon as possible.

Final Words
The FLC has a clear mandate to work with the Assistive Technology community and we welcome both the opportunities and challenges that are ahead. I want to thank you for this opportunity to speak here today and commend you madam Chair and members of the Subcommittee for your interest and leadership on this increasingly important subject.

Federal Laboratory Consortium for Technology Transfer

Vision

To maximize the return on investment from the nation's research and development

Mission

To add value to the Federal agencies, laboratories, and their partners to accomplish the rapid integration of R&D resources within the mainstream of the U.S. economy

Strategic Goals & Objectives

Enhance Communication

- ★ Expand communication among member agencies and their laboratories
- ★ Increase dialogue with State and local governments, businesses, academia and other external participants
- ★ Publicize best practices, solutions, and success stories

Leverage R&D Investments

- ★ Explore innovative approaches to technical assistance and other technology transfer activities
- ★ Reduce time, cost and risk of R&D projects
- ★ Increase cost sharing collaborations
- ★ Increase use of Federal technology by all participants

Improve and Innovate Technology Transfer Process

- ★ Characterize and analyze agency technology transfer policy, procedures, and activities
- ★ Address barriers identified by external participants and others
- ★ Provide fundamental and advanced education and training to enhance the technology transfer profession
- ★ Provide to Federal agencies analysis of key performance measurement elements and assessment options

Federal Laboratory Consortium Assets

- ★ Statutorychartered network of agencies including over 600 laboratories and technical facilities
- ★ Focal point for timely internal and external access to Federal R&D activities
- ★ Broad based forum for technology transfer issues resolution
- ★ Cost effective mechanism for training, outreach, and partnering
- ★ Proven regional networks and resources
- ★ Objectivity, credibility, and outstanding performance recognized and accepted by stakeholders

Figure 1

Federal Laboratory Consortium
Statutory Mandates

Develop and administer techniques, training courses, and materials concerning technology transfer to increase awareness of Federal laboratory employees regarding the commercial potential of laboratory technology and innovations

Furnish advice and assistance to Federal agencies and laboratories for use in their technology transfer programs

Provide a clearinghouse for requests for technical assistance from States and units of local governments, businesses, industrial development organizations, and not-for-profit organizations including: universities, Federal agencies and laboratories

Facilitate communication and coordination between Offices of Research and Technology Applications of Federal laboratories

Utilize the expertise and services of Federal agencies for technology transfer

Facilitate the use of appropriate technology transfer mechanisms

Assist Federal laboratories to establish programs using technical volunteers to provide technical assistance

Facilitate communications and cooperation between Offices of Research and Technology Applications of Federal laboratories and regional, State, and local technology transfer organizations

Assist colleges or universities, businesses, not-for-profit organizations, State or local governments, or regional organizations to establish programs to stimulate research and to encourage technology transfer

Seek advice in each Federal laboratory consortium region from representatives of State and local governments, large and small business, universities, and other appropriate persons on the effectiveness of the program

Figure 1

Federal Laboratory Consortium

Strategic Focus Areas

- **Creating Innovative Partnerships:** we are *listening* to industry, interacting with trade associations on a number of levels and responding to their specialized technological needs.

- **Influencing Technology Policy:** we are capitalizing on our experience and expertise in technology transfer to clarify the issues effectively and influence the science and technology policy debate.

- **Optimizing our Diverse Resources:** we are coordinating our various interagency efforts to develop improved strategies and opportunities for moving government technologies to the market.

- **Strengthening the FLC Structure:** as we provide the forum for agencies to collaborate, we are "reinventing" the FLC to match the new emerging technology needs of the 21st century.

- **Leading the Vision:** reflecting the new FLC, we are sharing information on partnering and experience with policy people to meet -- to anticipate -- the demands of changing inquiries and resources to make the most of Federal technology, as we head into the next century.

- **Projecting a Positive and Consistent Image:** while continuing to develop strong industry-Federal partnerships, we take every opportunity to raise the awareness of the successful technology transfer between laboratories and industry and the breadth and depth of the FLC as a resource, and to explain and publicize our mission and services.

Figure 2

FEDERAL LABORATORY CONSORTIUM
OFFICE OF THE WASHINGTON, DC REPRESENTATIVE

ASSISTIVE TECHNOLOGY ACTIVITIES

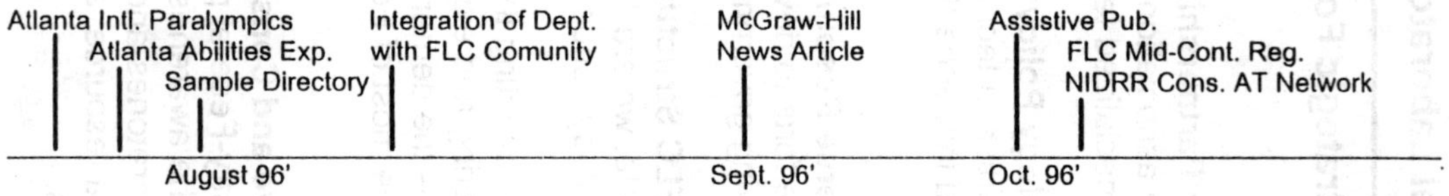

ataclv1.doc

Figure 3

FEDERAL LABORATORY CONSORTIUM
OFFICE OF THE WASHINGTON, DC REPRESENTATIVE

ASSISTIVE TECHNOLOGY ACTIVITIES

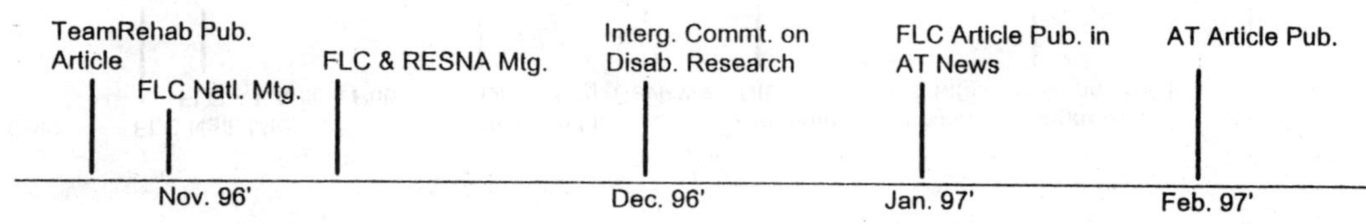

atactiv2.doc

Figure 3

FEDERAL LABORATORY CONSORTIUM
OFFICE OF THE WASHINGTON, DC REPRESENTATIVE

ASSISTIVE TECHNOLOGY ACTIVITIES

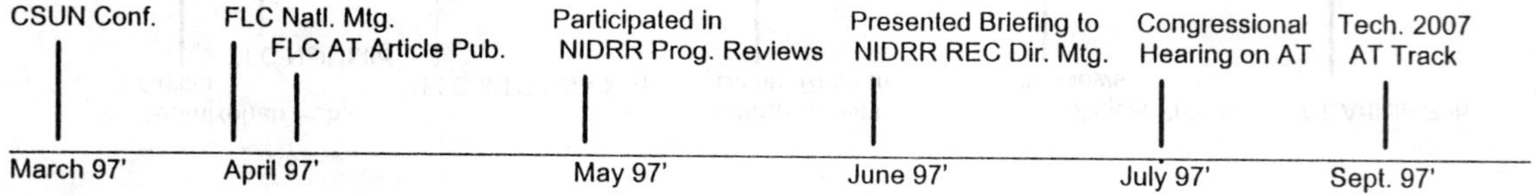

atacliv3.doc

Figure 3

Federal Laboratory Consortium

The FLC Mission:

To promote and facilitate the rapid integration of federal laboratory expertise, research and technologies into the mainstream of the U.S. economy.

FLC MEMBER AGENCIES

Central Intelligence Agency
Department of Agriculture
Department of the Army
Department of the Air Force
Department of Commerce
Department of Defense
Department of Education
Department of Energy
Department of Health and Human Services
Department of the Interior
Department of Justice
Department of the Navy
Department of Transportation
Department of Veteran's Affairs
Environmental Protection Agency
National Aeronautics and Space Administration
National Science Foundation
Tennessee Valley Authority

C. Dan Brand, M.P.H.
Associate Director for Technology Advancement
HHS/FDA/National Center for Toxicological Research
Jefferson, Arkansas

Dan Brand is currently the Chairman of the Federal Laboratory
Consortium and has served as the FLC Representative for the National
Center for Toxicological Research (NCTR) since 1992. Prior to being
elected FLC Chairman, he served as the Mid-Continent Regional
Coordinator (95/96) and as a member of the Executive Committee and
Board of Directors. As the Regional Coordinator he initiated several
projects that are now being replicated across the FLC network. His
leadership with the Mid-Continent Regional Technology Transfer Center
has resulted in technology partnerships from the labs to businesses
across the 14-state region. He was responsible for establishing FLC
program support with the Small Business Administration and the Small
Business Development Center. This collaboration has enhanced the
effectiveness of Technology Transfer across the region. He is the FLC
organizer and liaison with the Biotechnology Industry Organization.
He is very active in marketing the FLC, as an organization and the
member labs technologies. Under this guidance, the "10 Steps to
Federal Technology Transfer" has been presented to over 1000 people
across the U.S. The region has a Webpage and roundtable for effective
communication within the region. His participation at national SBIR
conferences has resulted in development of CRADAs and license
agreements. He started the Biotechnology round table on the FLC
Zyncom. He has served as the Chairperson for the IGER, Chairperson
for the Mid-Continent Outreach Committee, a member of the Small
Business Committee and a member of the `94 Spring Program Committee.
He is a regular presenter of FLC & Technology Transfer to the OPM
training courses and serves on the FLC Policy and Planning Committee.
Dan has received the Mid-Continent Representative of the Year Award.
Development of the "TAP-IN to SUCCESS Stories" has received excellent
review from both industry and labs. He is currently directing a
research project to provide lab directors and ORTAs "Best Practices"
in technology transfer to be used as a model for the Nation.

Mr. Brand has a broad range of experience in
government/industry/academia interactions. He has served as a
scientist and Senior Level Manager for FDA in St. Louis and Kansas
City, Missouri; Nashville, Tennessee; Washington, DC; and Jefferson,
Arkansas. He has personally managed 230 employees and a yearly budget
of $20 million. During a two-year period, he was responsible for
increasing productivity by 86%, while cutting costs by $2 million. In

addition to his many awards at the FDA, he has received commendations for exceptional performance by Governors of New Mexico, Nebraska, Iowa, and Arkansas. He has served as President of the Senior Executive Association Chapter and President of a local Chamber of Commerce. He is a graduate of the OPM Federal Executives Institute. He has over 70 publications and has made 400 public presentations before professional audiences, including congressional hearings. He is a founding member and serves on the Board of Directors of the Arkansas Biotechnology Association. He is the founder and current President of the Arkansas Chapter of the Technology Transfer Society and was recently elected to the National Technology Transfer Society Board of Directors for 96/98. He serves on the Arkansas Governor's Science, Research, and Development Planning Committee. He serves on the Arkansas State University Alumni Board of Directors and chairs the Scholarship Committee. He is an instructor in Technology Transfer for the University of Arkansas at Pine Bluff, and has developed training in technology transfer for other organizations.

Mrs. MORELLA. Thank you, Mr. Brand. And thank you also for your written testimony, which gives a number of examples of what some of our federal agencies are doing with regard to assistive technologies, and the various charts that accompany it.

I'd now like to recognize Dr. Bruce Webbon from NASA.

TESTIMONY OF BRUCE WEBBON, CHIEF, COMMERCIAL TECHNOLOGY, AMES RESEARCH CENTER, NATIONAL AERONAUTICS AND SPACE ADMINISTRATION, MOFFET FIELD, CA

Mr. WEBBON. I'd also like to thank the Committee for the invitation to speak here today. I want to preface my remarks by saying that I was asked to speak from my experience as a federal employee in a field center, actually responsible for developing some of these devices in the past, and to try and tell you a couple of stories about some of the things that we have done, and to try and draw some conclusions about what enabled those things to work, and perhaps what obstacles there were, and perhaps some things that might be done to change that and make it easier in the future.

I'm not representing NASA headquarters or NASA policy. So it's really just anecdotes from the field.

Most of my experience has resulted from requests from patients, physicians or small companies who have approached NASA. They had particular needs and they thought that perhaps we could fill those needs with some of the technology that was developed for our primary mission.

It was not the result of a focused outreach effort in the past. And NASA and other government agencies are moving in that direction, and I'm excited by prospect.

I think we have made a lot of contributions in the past, and I think that with a more focused effort, we can do a lot more.

As you certainly know, of course, the number of citizens with disabilities is large, and it is increasing. And the kinds of devices that I have experience with, the results from the individual inquires, their needs are often very particular to that person.

So there's not a very large market, and as a result, companies don't invest the resources that it would take to develop these specialized devices. So there really isn't anywhere else for people to go if they don't know how to do the things themselves.

The technology may exist in federal labs, but the private sector is often not able to access it, and to actually get it in place to the citizens who actually need it. Agencies such as NASA and others certainly could contribute a lot, if they were directed to do so.

There are some specific issues that we need to address to make that happen. Now, we often have the necessary technical expertise, but the work is not within the scope of our primary mission.

And that can change with direction from the top. So when requests come in, they often are not serviced because there just is not enough time and enough resources to actually do what needs to be done.

It doesn't necessarily take new money or new resources, but it does take direction to satisfy some of these requests. We also have recognized in the past that when we build a prototype device, we can't just transfer it.

That's a word that needs some definition. You can't just throw it over the wall. If you're going to be successful and not only meet the needs of the individual citizen who had the request in the first place, but to see that the device ends up as perhaps a low production device in the commercial market, you have to follow through.

So there's a need for resources afterwards to work with the companies to collaborate and make sure that they really understand what needs to be done to see that the transfer is successful.

There are serious medical and patient liability issues that have to be addressed. When we get a request from a patient, we have to go through a formal procedure to make sure the liability and the product liability and human use and all those requirements are satisfied.

And those requirements, in fact, are getting more difficult for folks like us to actually work with. The system is getting more complex, not less, and that's often a barrier to doing the kinds of things we'd like to do.

In spite of that, the government and its labs have unique qualifications to do this kind of stuff. We often have unique technical expertise which has been developed over the years.

I can speak for NASA in my particular case, where our job was to build protective systems for hazardous environments and life support systems for use in hazardous environments. Those are things that are used directly by people.

So we had a lot of expertise in engineering and science and physiology and so forth. That expertise is developed to do the primary mission of the lab, but it often is then available for application for these other assistive devices, given the will to do it.

The labs, like Ames Research Center, also have the resources to fabricate and to test devices and to do that at fairly low cost, right in-house, and without a lot of money. So we have some unique credentials to do that.

NASA has been doing this for as long as it's existed. There are many devices that are currently on the market now that resulted from NASA technology.

The primary headquarters organizations responsible for this are the Life Sciences Division and the Office of Life and Microgravity Sciences and Applications, and the Commercial Technology Division of the Office of Aeronautics and Space Transportation Technology.

Recently, the Commercial Technology Division was given primary responsibility for ensuring that NASA technology is transferred to the private sector, and they have organized a program in the area of assistive devices. We're in the process of trying to put together a program plan to make that happen.

Their intent is to make NASA a model for how this kind of thing, this technology transfer, can be done in the future. This involves all NASA Centers.

So what I'll talk about just briefly, some of these examples, are of my particular experience, but all NASA Centers have similar stories.

The particular things I want to mention here are two programs, one in the past and one that's still current. They are different in the sense that one was the result of a focused programmatic effort to actually go and make a specific thing happen, and the other was a case of just serendipity, where a request came in and we met the needs of that request, and devices are now on the market as a result.

The programmatic case is the case of using cooling systems, liquid cooling technology, which was developed originally by the British for use by pilots, and was adapted by NASA for cooling astronauts while they're wearing space suits. And Ames Research Center had the job of investigating physiology, so basically doing research on those systems and how they worked on the human, and how we might improve the hardware.

That was in the early 1970's. We recognized very quickly that there were many applications, both industrial, military, race cars, as well as medical. So, for the last 25 years, we, in fact, have worked on many medical applications of cooling system technology.

In particular, a few years ago, the NASA Administrator signed a Memorandum of Understanding with a nonprofit organization, the Multiple Sclerosis Association of America. The intent of that MOU was to improve the technology for use by MS patients, and to direct us to actually do that work.

We started that with a workshop, and that's one of the points I want to make, is that while federal labs may have the technical experience and the resources, they often do not understand the problem.

So a key point, I think, in these kinds of programs is, you have to understand the requirements of the market, the requirements of the end users, and that's how that program began. It's been very successful.

NASA, the Life Sciences Division and Commercial Technology Division, have supported it for 3 years. We now have signed an MOU with both Lockheed Martin Corporation and NASA to bring private money in to continue the work.

So we're very proud of that program. We think it's the way things should be done.

Another program which resulted from just the target of opportunity is now called Medical Anti-Shock trousers. That began with a request from a physician at Stanford.

They had a patient who was hemorrhaging after surgery and they weren't able to stop it. And what was done was a G-suit, a pilot's G-suit was modified in just a few hours.

We took it over to Stanford and put it on the patient and it stopped the bleeding. We couldn't do that today.

That was about 20 years ago. We have too many obstacles today to actually do hands-on work, touching patients. We wouldn't be allowed to do that, but at that time we could.

The result since then is that there are about eight companies now making those devices, and that technology was adopted by the U.S. Space Foundation into the Space Technology Hall of Fame for technology transfer.

In their research into it, they came back and told us that it had been used more than two million times. It's only used—it's an emergency device and it's only used for people who are bleeding severely. A lot of those folks might not be around today if it hadn't been for that work.

So that was a target of opportunity. Another thing that we're doing is, we're trying to set up a volunteer program, and that would allow access by individuals who have disabilities to access Ames staff, and hopefully that will become a NASA-wide program to come up with specialized devices to meet their particular needs.

In this case, the technology is there; it exists in the form of the staff. The barriers to making it happen are the legal barriers that have to do with allowing people to come in.

We're going to build devices for them, and those devices will be used by people, so we have to cover those issues. workers comp if the people are hurt on the job, and so on. We're putting that in place, and Lockheed Martin, we hope, is also going to adopt that program.

In summary, to get through this very quickly, these examples, which are detailed more in the written testimony, show what might be done with the—what has been done in the past without a formal, organized program.

We think we can do a lot more with that direction coming from the top. Every NASA Center, I know, has similar stories. I'm aware of many of them. There's just not time to tell you about them.

The resources are often available so that new money and new resources do not necessarily need to be provided, but direction does have to be provided because these efforts, in my experience, are in addition to your primary job. So that needs to come from the very top to allow you to do it.

The Commercial Technology Division at NASA headquarters is in the process of putting together a plan to try and enable this to happen. Of course, the Life Sciences Division is also actively involved.

Thank you.

[The prepared statement of Mr. Webbon follows:]

Statement of
Dr. Bruce Webbon
Chief, Commercial Technology Office
NASA-Ames Research Center

Before the

Subcommittee on Technology
Committee on Science
U.S. House of Representatives

July 15, 1997

Madam Chair and Members of the Subcommittee, I am pleased to have the opportunity to discuss NASA technology and its application to assist people with disabilities.

Introduction

A significant and growing percentage of US citizens are afflicted with some form of physical disability. These range from relatively slight disabilities, which still may seriously interfere with their quality of life and work productivity, up to the near total disability caused by conditions such as severe spinal cord injury. In many cases technical solutions may exist which could benefit these citizens but that technology is often not available to them. In spite of the large overall number of citizens with some form of disability, the needs of particular individuals are often very dissimilar and as a result, the market for specialized assistive devices is often very small. As a consequence of this fragmented market, large private companies that may have sufficient scientific and engineering resources to develop the needed devices, are often reluctant to do so since the return on investment would be small or even totally non-existent. Small companies that might serve these niche markets often don't have the resources necessary to develop such new products. The end result is that while the necessary technology may exist in many cases, the private sector is often not able to apply it to serve the needs of many disabled citizens. The Federal Laboratories, including agencies such as the National Aeronautics and Space Administration, may be able to help to bridge this technology gap.

The following are some of the specific issues that must be addressed in order for this to happen:

- Many worthwhile applications of available technology to the problems of the disabled still require significant engineering and scientific effort to develop and demonstrate a prototype suitable for transfer to the private sector.

- The number of patients (potential customers) is often limited so that the return on private investment is often unacceptably low.

- Federal labs may have the necessary technical expertise but such work is often not within the scope of their primary mission.

- Resources, both staff time and dollars, are required and these are becoming increasingly scarce in the Federal Government.

- Prototype devices can not just be "thrown over the wall". Resources must be provided for follow-on support and collaboration with industry to ensure successful technology transfer.

- Serious medical and patient liability issues must be addressed during the development of both prototypes and actual commercial devices.

The Federal Government and its laboratories are uniquely qualified to address these issues and make major contributions in the area of medical assistive devices if they are given specific direction to do so as well as the resources to meet the challenge. Their most important resource is the technical expertise of the lab staffs and the facilities and specialized equipment that have often been built up over many years to accomplish the lab's primary mission. These resources are not tied to a company's "bottom line" so that a portion of their efforts can be re-directed if desired. Since they are not in the business of selling commercial products they have no commercial vested interest in any particular design solutions. Therefore, they can potentially be completely objective in selecting the best solution to meet the patient's needs. Most Federal labs also have the capability to design, fabricate, and test prototypes in-house at low cost and many labs already have the required legal review mechanisms in place to deal with the medical and patient liability issues.

<u>Overall NASA Efforts</u>

The National Aeronautics and Space Administration has a long history of applying the aerospace technology that has been developed to accomplish its primary mission to help to solve problems on Earth. Many medical devices that are currently on the market had their origins in NASA technology. The primary NASA headquarters organizations that have directed these field center efforts are the Life Sciences Division of the Office of Life & Microgravity Sciences & Applications and the Commercial Technology Division of the Office of Aeronautics & Space Transportation Technology. The Commercial Technology Division has been given primary responsibility for ensuring that NASA technology is expeditiously transferred to the private sector. Many examples of such technology transfer are reported in publications such as Spin-off Magazine and NASA Tech Briefs.

The Director of the Commercial Technology Division has already assigned staff to participate in the Interagency Committee on Disability Research, Subcommittee on Technology. This committee will produce a "Compendium of Federal Rehabilitation Technology Research." The Commercial Technology Division is also in the process of formulating a program plan that will focus resources on the problems of the disabled with the intent that NASA will become a model for the effective transfer and commercialization of technology to benefit people with disabilities. This program will utilize both Federal and private assets as appropriate in order to accomplish its goals. All NASA centers are involved in this new program just as all centers have made contributions to medical technology in the past. Some recent examples of NASA contributions to medical device technology include:

- Compliant cable mechanisms for wheel chair suspension

- Functional electrical stimulation (in collaboration with the Department of Veterans Affairs) for motor function restoration in persons with paralysis

- Exercise aid for walking rehabilitation (in collaboration with the Department of Veterans Affairs)

- Cooling therapy devices for use by persons with multiple sclerosis

- Filter technology for ambient eye tracker interference

- Collaboration with UCLA Brain Research Institute for spinal cord injury rehabilitation and repair

- Video game neurotherapy for attention deficit disorder

- Orthotic locking knee brace and prosthetic elbow joint

Specific Examples of Medical Assistive Device Development

I would like to give several specific examples based on my own experience as to how the process of technology transfer for medical assistive devices has worked at NASA's Ames Research Center. These will illustrate how the process has worked in the past and will point out how it might be improved in the future.

For many years NASA Ames was responsible for the development of advanced technology for space suits and portable life support systems for human space exploration missions. As a result of this primary mission, a highly skilled technical staff had been assembled along with the unique facilities and equipment they required. This technical area required staff expertise in a wide range of disciplines including various engineering skills as well as physiology and bio-mechanics. Since their role was advanced R&D, the staff had also developed a number of collaborations with both academia and industry. These staff skills were particularly applicable to the development of medical devices and over many years this group responded to requests from private companies, physicians, and individuals to assist in finding solutions to medical device problems. These efforts were enabled by a supportive and tolerant center management and in particular by the active support and participation of the Ames Chief Medical Officer who was responsible for ensuring that all human use and patient liability issues were properly handled.

As a result this group was able to develop prototypes of the following devices, many of which are currently being manufactured by commercial companies, in addition to accomplishing its primary job of providing advanced technology for NASA missions:

- suit to control bleeding in child hemophiliacs
- radiation cancer therapy helmet
- portable cooling system for workers in hot environments *
- device to assist with "patterning" therapy in brain damaged children
- device to assist senior citizens in rising from a chair
- head cooling systems for chemotherapy *
- cooling systems for pilots and race car drivers *
- "cool bra" for breast cancer screening
- cooling systems for children with erythromelalgia
- cooling systems for para and quadriplegics and MS patients *
- negative pressure therapy chamber for treatment of pneumocystis carinii
- circumferential pneumatic counterpressure system for control of hemorrhage and shock *
- cooling systems for military applications *
- USAF advanced pressure suits
- spinal cord injury patient immobilization system *
- thermal control systems for surgery patients *
- hard suit design *

*Those indicated by an * have resulted in successful commercial products*

These development efforts fall into two distinct classes, those that resulted from a planned, programmatic effort and those which occurred by serendipity due to the existence and reputation of the technical group. I will provide examples of both.

Liquid cooling garment systems were first conceived by the British Royal Aircraft Establishment during the 1950's to help aircraft pilots maintain their body temperature while wearing bulky protective suits in a hot aircraft environment. The concept was adopted and further developed by NASA during the Apollo program when it was found that astronauts wearing space suits suffered from similar heat stress. Both NASA and Russian astronauts still use refined versions of these cooling suits today.

Ames Research Center staff began research in the late 1960's to better understand the alterations in the human physiology caused by the use of such garments and then to apply that knowledge to the development of improved garments. They soon realized that there were many Earth applications of this technology that could be used in both hot and cold environments as well as for medical applications. The first focused effort was instigated by the U.S Bureau of Mines which asked us to develop a portable cooling system that could be used by mine rescue workers who were required to work very hard in very hot environments following disasters such as a fire in a mine. This effort resulted in the first prototype of the portable cooling systems that are currently produced by a number of companies.

In 1978 a physician at UCLA, who knew that the symptoms of multiple sclerosis patients were greatly exacerbated in a hot environment, approached us to request that we try to use an astronaut type cooling system on some of his patients to determine if artificial cooling would provide any symptomatic relief. We took an experimental cooling system to a physical therapy clinic at UCLA and determined that it did indeed appear to provide some relief to the patients. A private company then adopted the system design and began to produce commercial systems for sale to MS patients.

These systems are still available today but they are not widely used in the MS community primarily because the scientific foundation establishing their effects has not been established and the currently available devices are expensive and simultaneously relatively expensive. In 1994 the NASA Administrator signed a Memorandum of Understanding with the Multiple Sclerosis Association of America and committed us to collaborate with the MS community in order to improve the technology and expand its use.

We recognized immediately that we did not understand the needs of the MS community. Therefore, we began the program by organizing an invited workshop that included members of the MS research and patient care community as well as representatives of the companies that were producing cooling systems for use by MS patients. The purpose of this workshop was to define the specific requirements of the program to which we had been committed by the Administrator. We believe that this is a critical first step in any such program. The needs and requirements must be identified and defined by the actual end users of the technology to be developed.

This program has been extremely successful. We determined during the workshop that the most critical problems that we could assist the MS community in solving were to collaborate with them in performing carefully documented scientific experiments to define and quantify the effects of cooling on MS patients. This would establish the scientific basis of cooling therapy. In addition the small companies involved would benefit greatly if NASA undertook the development of several key hardware components that could be used to improve their devices. We have largely accomplished the first goal and the result is a number of scientific papers that validate the efficacy of cooling therapy for MS patients. As a result of these publications the technique is being widely adopted including the establishment of cooling therapy clinics in Veterans Affairs' hospitals and elsewhere. We collaborated with a university to develop the hardware components. This was accomplished by teams of undergraduate students working under a cooperative agreement with NASA so that the cost was very low. The student teams produced several innovative designs that will result in patents in addition to the valuable real world experience they obtained. These component designs will be made available to the manufacturers in the near future.

This program, which has been supported by both the Life Sciences Division of the Office of Life & Microgravity Sciences & Applications and the Commercial Technology Division of the Office of Aeronautics & Space Transportation Technology over the last 3 years, was not completed during the planned 3 years. Therefore, we have recently brought in a large corporate partner, the Lockheed Martin Corporation to help to provide the resources necessary to complete the program during FY98. In order to accomplish this a Memorandum of Understanding was negotiated and signed in June of this year between NASA Ames Research Center, Lockheed Martin Corporation, and the Multiple Sclerosis Association of America. Each of the participants has pledged to contribute both financial and other resources necessary to complete the program. We are extremely proud of this program and we believe it can serve as a model of the methodology to be used for similar programs in the future.

The development of such applications of liquid cooling technology, including this MS application, was selected by the United States Space Foundation in 1993 to be inducted into the Space Technology Hall of Fame.

The second detailed example, which was also accomplished by the Ames Research Center staff, will illustrate how an informal effort can produce highly significant results if it is nurtured by lab management. This example is the development of what has become known as Medical Anti-Shock Trousers (MAST).

Special suits, which help to protect the pilots of high performance aircraft from passing out during violent maneuvers, have been available for many years. These "anti-g" suits work by applying external pressure to the legs and abdomen to help to maintain blood pressure in the trunk and more importantly in the brain. In the early 1950's such a suit was used to attempt to maintain blood pressure during cranial surgery but the technique was not widely accepted. The US military also investigated the use of such suits during surgery. The Ames staff were actively involved in the development and testing of these suits and were aware of such applications.

In 1969 a surgeon from the Stanford University Hospital placed an emergency call to an Ames space suit physiologist asking for help. Stanford had a female patient who was dying due to uncontrollable internal bleeding following surgery. Several surgical attempts had already been made to stop the bleeding but they were not successful. The surgeon reasoned that someone at Ames might be able to help and fortunately he was able to reach the right person. A small team of staff members were able to modify a pilot's g-suit and assemble a pneumatic controller within a few hours. This system was taken to the hospital and used on the patient. The bleeding stopped within a few hours and the patient recovered. A paper documenting this technique, "Anti-G Suit as a Therapeutic Device" was published in the Journal of Aerospace Medicine, Vol.41, No.8, August 1970.

Over the next several years the Ames staff responded to several similar calls still using borrowed and adapted hardware. In 1975 several Ames engineers collaborated with the Ames medical officer to design and fabricate a suit and controller especially for this purpose. This system was then used a number of addional times and the results were documented in a survey paper reviewing the use of the technique which was published in the Journal of the American Medical Association; Feb 16,1979-Vol 241, No.7. The editorial in that same issue advocated that trauma physicians consider the use of this life saving technique. As a result of this advocacy, a number of companies began to manufacture anti-shock trouser systems.

These systems are now in common use in ambulances and emergency rooms around the world. The development of these anti-shock trousers was selected by the United States Space Foundation in 1996 to be inducted into the Space Technology Hall of Fame. As part of the Space Foundation's due diligence research they determined that these life saving systems have been used more than 2 million times over the last 20 years.

This example illustrates what can be accomplished without a formal program by a knowledgeable and dedicated staff provided their management supports and nurtures such efforts.

A final example will illustrate how government staff expertise and resources might be utilized on a volunteer basis to meet the needs of individual citizens with disabilities.

Ames Research Center is currently working with a non-profit organization called the Tetra Society to define and implement the necessary procedures to enable Ames staff to work directly with disabled individuals. The Tetra Society acts as a clearing house to connect disabled individuals directly with engineers and technicians who can produce specialized devices that will improve their quality of life. For example, the specific disability of paraplegics varies greatly. Many have individual needs for devices allowing them to reach a telephone for example or to carry a back pack on the back of their wheel chair and then be able to move it within reach. These devices can make significant improvements to their quality of life and self-sufficiency but they must often be custom made for each individual. Tetra attempts to match such individuals with the technically skilled people who can fabricate devices to meet their specialized needs.

NASA is a gold mine of such people but they can not be accessed by the disabled community unless a number of legal and liability hurdles are crossed. These include human use issues, Worker's Compensation, and product liability. With the enthusiastic support of center management, the Ames Commercial Technology Office Staff is currently in the process of establishing legal and administrative procedures that will enable technical staff members to volunteer to use both their expertise as well as specialized fabrication and test equipment available at the center to fabricate such devices. We hope that this will become a model for other NASA centers as well as private companies. The Lockheed Martin Corporation has already indicated that they would like to set up a similar volunteer mechanism following our lead.

These specific examples illustrate what may be accomplished even without a formally organized, Agency-wide program. We know that every NASA center has many similar and often unpublicized examples of providing assistance to the disabled community. We believe that far more can be accomplished given a more focused and organized Agency-wide approach.

Summary

Agencies such as NASA have already made significant contributions to the disabled community: Both the Commercial Technology Division of the Office of Aeronautics & Space Transportation Technology and the Life Sciences Division of the Office of Life & Microgravity Sciences & Applications have supported and will continue to support such applications. Because of limited resources available to NASA, only a limited number of such applications may be pursued at any one time. As a technology becomes commercially available and the research to describe its use and scientific merit is published or transferred to the NIH or other user agency, the activity is phased out to release resources for another application.

So-called technology transfer in actuality means the transfer of know-how together with making the technology commercially available. Therefore, the most important NASA asset is the technical strength of its staff. This must be preserved above all else to enable such contributions in the future.

Bruce W. Webbon, Ph.D.

Curriculum Vitae

June 1997

Date and Place of Birth: June 7, 1945; Bridgeport, Connecticut

Education:

School	Dates	Major	Degrees and Date Awarded
Georgia Institute of Technology	Sept. 1965 - Sept. 1967	Mechanical Engineering	B.S. 1967
University of Florida	Sept. 1968 - Sept. 1969	Mechanical Engineering	M.S. 1969
University of Missouri	Sept. 1972 - June 1978	Mech. & Aero. Eng.	Ph.D. 1978

Master's Thesis: *The Physical Mechanisms of AC Conductor Corona*, University of Florida (1969)

Ph.D. Dissertation: *Nucleate Pool Boiling of Water on Arrays of Artificial Sites in Normal and Reduced Gravity Environments*, University of Missouri (1978)

Additional Training:

United States Air Force (Beale AFB)	- USAF Tankers, Trainers, and Bombers Refresher Training (1993)
Hyperbarics International	- Undersea & Hyperbaric Medical Training Program (1992)
University of Southern California	- Hyperbaric Physiology and Diving Accident Emergency Management (1989)
NASA-Johnson Space Center	- Hypobaric Physiology Training (1989)
National Association of Underwater Instructors (NAUI)	- SCUBA Instructor (1988)
Stanford University	- Courses in mathematics and bioengineering (1976, 1978)
American Institute for Professional Education	- Microprocessor fundamentals (1978)

Military Service: None

Security Clearance: Secret

Professional Experience:

August 1996 to present	***Chief, Commercial Technology Office*** NASA Ames Research Center Responsible for managing all technology transfer and commercialization activities for Ames Research Center. The Commercial Technology Office, which reports to the Center Director, has a professional staff of approximately 30 people and an annual budget of approximately $3.5M.
March 1990 to August 1996	***Branch Chief***, Extravehicular & Protective Systems Branch Space Technology Division NASA Ames Research Center Responsible for managing the activities of a professional staff consisting of approximately 25 people focused on basic and applied research and development of protective systems to allow humans to work safely in hostile environments. The staff includes 4 Ph.D.s, 7 M.S.s, and 5 B.S.s in disciplines such as engineering, psychology, and physiology. Annual Budget approximately $1.5M
1995 to present	***Instructor*** LaunchSpace, Inc. Instructor for short course entitled: Life Support and Protective Systems Design.
1993 to present	***Adjunct Professor*** Stanford University Instructor for HB 107, Biology and Space Exploration.
January 1988 to March 1990	***Branch Chief***, Crew Research and Space Human Factors Branch Aerospace Human Factors Research Division NASA Ames Research Center Responsible for the planning and initiation of a new research organization. Responsible for the activities of a professional staff consisting of approximately 25 people including 6 Ph.D.s, 1 M.D., 8 M.S.s, and 5 B.S.s in disciplines such as engineering, psychology, and physiology. Branch activities are devoted to basic and applied research and systems development in the areas of space and aeronautical human factors and protective systems technology for both space and terrestrial applications.
October 1987 to October 1992	***Program Manager***, EVA/Suit Element of NASA's Pathfinder Program NASA Ames Research Center Manager of a program to develop advanced systems for solar system exploration.
October 1986 to January 1988	***Research Scientist***, Man-Vehicle Systems Research Division NASA Ames Research Center Responsible for a variety of programs to develop advanced technology for portable life support systems, protective suits, and advanced extravehicular work systems.
July 1984 to October 1986	***Partner***, Lakeview Research Aerospace consultants in life support systems and human engineering.

Bruce W. Webbon, Ph.D.

Professional Experience *continued*

| April 1983 to
August 1986 | *Program Director*, Bioinstrumentation Technology
SRI International |

April 1983 to
August 1986

Program Director, Bioinstrumentation Technology
SRI International

Program Director for biomedical instrumentation technology. Manager of a professional staff of approximately 15 people including 4 Ph.D.s, 4 M.S.s, and 3 B.S.s in disciplines such as electrical and mechanical engineering and physiology. Engaged in research and development of state-of-the-art physiological instrumentation systems for space, military, and commercial applications.

June 1982 to
April 1983

Program Manager, Spaceplane Program
SRI International

Manager of the "Spaceplane Concept Feasibility Evaluation Program" to perform design and systems analysis of a high performance, manned, military space vehicle.

November 1981 to
June 1982

Senior Research Engineer,
SRI International

Responsible for development of programs in life support and protective systems for space and other environments. This includes basic research into physical/chemical processes applicable to protective systems.

March 1975 to
October 1981

Research Scientist,
NASA Ames Research Center

Responsible for development of portable life support system technology. This involved management of NASA, university, and contracted research and component development. Research activities included: 0-g aircraft flight experiments in boiling heat transfer; development of advanced, integrated space suit/portable life support systems; studies of human heat and mass balance during exercise; application of this technology to medical research and equipment.

August 1972 to
March 1975

Research and Teaching Assistant,
University of Missouri

Research:
  • Dissertation research in boiling heat transfer in 0-gravity
  • Computer model of 1-dimensional solid/liquid phase change process
  • Evaluation of gas chromatography for operating room use
  • Participant in Medical/Engineering working groups
Teaching:
  • Prepared and taught a senior/graduate course: "Life Support Systems Design"

September 1969 to
August 1972

Senior Thermodynamics Engineer,
Vought Systems Division of LTV Aerospace Corporation

Responsibilities included computer simulation of environmental control and life support systems; systems studies of advanced extravehicular protective systems; extravehicular systems studies for the space shuttle; experimental and analytical studies of advanced heat sinks including fusible, space radiators, and flash evaporators; and technical proposal team member. Served as Apollo Program Thermal Analysis Coordinator for thermal analyses performed at NASA/Johnson Space Center.

Professional Experience *continued*

September 1968 to September 1969	*Research Assistant in Mechanical Engineering,* University of Florida • Heat transfer and fluid mixing experiments funded by NASA grant • Thesis research on atmospheric corona discharge • Research involving similarity solutions to partial differential equations
September 1967 to September 1968	*Turbine Aerodynamics Analysis and Test Engineer,* Pratt and Whitney Aircraft Responsible for the design, fabrication, and use of a turbine aerodynamics test facility.

Professional Societies and Activities:

Member:
Professional Societies
- American Association for the Advancement of Science
- Aerospace Medical Association
- American Institute of Aeronautics and Astronautics
- National Association of Underwater Instructors

Member:
Committees
- US/USSR Joint Working Group on Space Biomedical and Life Support Systems (1990)
- AIAA Life Sciences and Systems Committee (1987-1990)
- USAF Advisory Committee on Military Man in Space (1988-1989)
- USAF Space Medicine/Human Engineering Panel (1985)
- National Research Council Committee on Chemical Protection (1979-81)

Chairman:
- EVA/Manned Systems Panel, NASA-OAST Workshop: "Technology for Space Station Evolution," January 1990
- EVA Session, 1984, 1985, 1987 Intersociety Conference on Environmental Systems
- Protective Systems Session, 1977 Intersociety Conference on Environmental Systems

Recent Invited Seminars / Teaching:
- Lecturer, Stanford University: Human Biology 107, Biology and Space Exploration (1993 to present)
- "Extravehicular Work Systems for Space Exploration Missions," Stanford University (February 1993)
- "Human Requirements for Life Support in Space," Institute of Electrical and Electronics Engineers: Engineering in Medicine and Biology Society (February 1993)
- "EVA Systems for Solar System Exploration" Stanford University/Asilomar (June 1992)
- "Human Requirements for Life Support in Space" Navy Postgraduate School (November 1988)
- "Human Requirements for Life Support in Space" University of California-Berkeley (January 1992)
- "EVA Systems for Space Exploration Missions," Massachusetts Institute of Technology (November 1991)
- "Human Factors in the Design of Space Suits and Life Support Systems," Human Factors Society 35th Annual Meeting (August 1991)
- "Extravehicular Work Systems for Space Exploration Missions," Visit to Mars Workshop, University of Utah (June 1990)
- Advanced technology for EVA," presentation to National Space Council Space Exploration Initiative Synthesis Team (August 1990)

Invited Participant:
- Conference on the Role of Heat Transfer in Medicine and Biology, University of Illinois (March 1976)
- International Colloquium on Drops and Bubbles, California Institute of Technology (August 1974)

Listed In:
- American Men and Women in Science
- Who's Who in the West

Inventions:

Patents

Tubular Sublimator/Evaporator Heat Sink, U.S. Patent No. 4,007,601
Space Suit Liquid Cooled/Vent Garment System, U.S. Patent No. 4,095,593
Space Suit Torso Closure, U.S. Patent No. 4,091,465
Pressure Suit Joint Analyzer, U.S. Patent No. 4,311,055
Quick Connect Coupling, U.S. Patent No. 5,392,844
Cooling Apparatus and Couplings Therefore, U.S. Patent No. 5,261,482

Invention Disclosures

Space Suit Carbon Dioxide Absorption System, NASA Tech Brief, No. 72-10168
A Portable, Fusible Heat Sink, disclosure filed, SRI Case No. P-1838
Nonventing Wine Bottle Cap, disclosure filed, SRI Case No. P-1851
A Method to Selectively Stimulate Individual Neurons within a Nerve Bundle Using Spatial and Temporal Superposition of Sub-threshold Electric Fields, disclosure in preparation
Lockable Bearings for Pressure Suit and Robotic Applications, disclosure in preparation

Honors and Awards:

- United States Space Foundation, 1996 Inductee into the Space Technology Hall of Fame for the development of Anti-Shock Trousers

- NASA Space Act Award for invention (1995)

- United States Space Foundation, 1993 Inductee into the Space Technology Hall of Fame for the development of Liquid Cooling Garment Systems

- National Research Council Post-Doctoral Research Advisor

- NASA Space Act Award for invention (1989)

- NASA award for outstanding management (1988)

- Society of Automotive Engineers, 1987 Award for Excellence in Oral Presentation

- 2 SRI awards for outstanding management (1985, 1986)

- 10 NASA awards for inventions (1974-1991)

- 8 NASA awards for achievement (1974-1982)

Journal Publications

1. B.W. Webbon, et al., "A Portable Personal Cooling System for Mine Rescue Operations," published in *Journal of Engineering for Industry* (February 1978).
2. B.W. Webbon, et al., "Comparison of Three Liquid/Ventilation Cooling Garments during Treadmill Exercise," published in *Aviation, Space and Environmental Medicine*, pp. 408-415 (July 1981).
3. R. Lantz and B.W. Webbon, "Measurement of Metabolic Responses to an Orbital, Extravehicular Work Simulation Exercise," 1988 SAE Transactions, *Journal of Aerospace.*
4. R. Williamson, et al., "Metabolic Responses to Simulated Extravehicular Activity," 1992 SAE Transactions, *Journal of Aerospace.*
5. D.J. Newman, H. Alexander, and B.W. Webbon, "Energetics and Mechanics for Partial Gravity Locomotion," *Aviation, Space and Environmental Medicine* (September 1994).
6. B.S. Yendler, B. Webbon, I. Podolski, and R.J. Buhal, "Capillary Motion of Liquid in Granular Beds in Microgravity," *Adv. Space Res.* Vol. 18, No. 4/5, 1996.
7. Y.A. Buyevich and B.W. Webbon, "Dynamics of Vapour Bubbles in Nucleate Boiling," accepted for publication, *International Journal of Heat and Mass Transfer*, Vol.39, No.12, pp 2409-2426.

Publications in Process

1. Y.T. Ku, L.D. Montgomery, and B. Webbon, "Hemodynamic and Thermal Responses to Head and Neck Cooling in Women," accepted for publication, *Aviation, Space, and Environmental Medicine.*
2. Y. Buyevich and B. Webbon, "Bubble Formation at a Submerged Orifice in Reduced Gravity," accepted for publication, *Chemical Engineering Science.*
3. Y. Buyevich and B. Webbon, "The Isolated Bubble Regime in Pool Nucleate Boilimng," accepted for publication, *Int. J. Heat Mass Transfer.*
4. Y. Buyevich and B. Webbon, "Towards a New Theory of Nucleate Pool Boiling," submitted for publication.
5. Y.E. Ku, L.D. Montgomery, B.W. Webbon, "Hemodynamic and Thermal Responses to Head and Neck Cooling in Men and Women," accepted for publication, *Am. J. Phys. Med. Rehabil.*
6. Y.E. Ku, L.D. Montgomery, K.C. Wenzel, and B.W. Webbon, "Physiological and Thermal Responses of Male and Female MS patients to Head and Neck Cooling," submitted for presentation at the 1996 Annual Meeting of the Consortium of Multiple Sclerosis (MS) Centers, September 26-28, 1996, Atlanta, GA.
7. L.D. Montgomery, Y.E. Ku, H.M. Hanish, and B.W. Webbon, "An Ambulatory Physiologic and Thermal Monitor for Study of Multiple Sclerosis Patients," submitted for presentation at the 1996 Annual Meeting of the Consortium of Multiple Sclerosis (MS) Centers, September 26-28, 1996, Atlanta, GA.

Conference Publications

1. J. Williams, B.W. Webbon, and R. Copeland, "Regenerable Thermal Control and Carbon Dioxide Control Techniques for Use in Advanced Protective Systems," published in *Proceedings of the 2nd Conference on Portable Life Support Systems*, NASA Special Publication SP-234 (May 1971).
2. B. Williams, G. McEwen, L. Montgomery, and B.W. Webbon, "Effectiveness of a Modular Liquid-Cooling Garment vs. a Standard Apollo Liquid Cooling Garment at Four Work Levels," presented at the Aerospace Medical Association Annual Scientific Meeting, May 1976.
3. H.C. Vykukal and B.W. Webbon, "High Pressure Protective Systems Technology," ASME Paper No. 79-ENASS-15 presented at Ninth Intersociety Conference on Environmental Systems, San Francisco, CA, July 1979.

4. B.W. Webbon, C.S. Weaver, and L.D. Montgomery, "Automatic Control of Liquid-Cooling Garment Inlet Temperature," published in *Proceedings of the Aerospace Medical Association 1983 Annual Scientific Meeting.*

5. B.W. Webbon and H.C. Vykukal, "EVA in the Space Station Era," ASME Paper No. 831135, presented at the 13th ICES (July 1983).

6. B.W. Webbon, "Ambulatory Physiological Monitoring Systems for Aerospace Applications," presented at Aerospace Medical Association 1984 Annual Scientific Meeting (abstract in *Aviation, Space and Environmental Medicine* (May 1984).

7. B.W. Webbon and H.C. Vykukal, "A Comparison of Space Suit Joint Flexure Forces as a Function of Suit Pressure," ASME Paper No. 840980 presented at 1984 Intersociety Conference on Environmental Systems.

8. B.W. Webbon and H.C. Vykukal, "A Method to Predict Metabolic Work Based on Measured Space Suit Joint Mechanical Work," presented at Aerospace Medical Association 1985 Annual Scientific Meeting.

9. J. Thompson, K. Brossel, and B.W. Webbon, "An Evaluation of Options to Satisfy Space Station EVA Requirements," Society of Automotive Engineers Paper No. 861008 (July 1986).

10. B. Squire and B.W. Webbon, "Development of a Thermal Control Coating for Space Suits," Society of Automotive Engineers Paper No. 871474 (July 1987).

11. R. Lantz, H. Vykukal, and B.W. Webbon, "An Innovative Exercise Method to Simulate Orbital EVA Work: Applications to PLSS Automatic Control," Society of Automotive Engineers Paper No. 871475 (July 1987).

12. C. Lomax and B. Webbon, *Direct Interface Fusible Heat Sink for Astronaut Cooling,* NASA TM 102835, May 1990 also SAE Paper No. 901433, International Conference on Environmental Systems, July 1990.

13. D. J. Newman and B.W. Webbon, "Human Locomotion and Workload for Extravehicular Activity (EVA): Simulated Partial Gravity Environments," published in *9th International Academy of Astronautics Man in Space Conference Proceedings.* (June 1991).

14. R. Williamson, P. Sharer, and B. Webbon, "Metabolic Responses to Simulated Extravehicular Activity," SAE Paper No. 921303, presented at 22nd International Conference on Environmental Control Systems, July 1992.

15. L. Smith, S. Nair, J. Miles, and B. Webbon, "Human Thermal Modeling for Portable Life Support System Control," SAE Paper No. 932186, presented at 23rd International Conference on Environmental Control Systems, July 1993.

16. B. Yendler and B. Webbon, "Capillary Movement of Liquid in Granular Beds," SAE Paper No. 932164, presented at 23rd International Conference on Environmental Systems, July 1993. 23. L. Smith, S. Nair, J. Miles, and B. Webbon, "Human Thermal Modeling for Portable Life Support System Control," SAE Paper No. 932186, presented at 23rd International Conference on Environmental Systems, July 1993.

17. R. Williamson, P. Sharer, and B. Webbon, "A Unique Exercise Device for Simulating Orbital Extravehicular Activity," presented at AEROTECH Space Simulation Conference, September 1993.

18. L. Smith, et al., "Issues in the Development of Automatic Thermal Control for Portable Life Support Systems," SAE Paper No. 941383, presented at 24th International Conference on Environmental Systems, Friedrichschafen, Germany, July 1994.

19. A. Campbell, et al., "Modeling the Sweat Regulation Mechanism," SAE Paper No. 941259, presented at 24th International Conference on Environmental Systems, Friedrichschafen, Germany, July 1994.

20. B. Webbon, B. Yendler, and J. Miles, "Nucleate Pool Boiling of Water in Normal and Reduced Gravity Environments," SAE Paper No. 941448, presented at 24th International Conference on Environmental Systems, Friedrichschafen, Germany, July 1994.

21. B. Yendler, et al., "Capillary Movement of Liquid in Granular Beds in Microgravity," SAE Paper No. 941449, presented at 24th International Conference on Environmental Systems, Friedrichschafen, Germany, July 1994.

22. X. Chalandon and B. Webbon, "Human Heat and Mass Transfer in Microgravity and Hypobaric Environments," SAE Paper No. 941318, presented at 24th International Conference on Environmental Systems, Friedrichschafen, Germany, July 1994.

23. A. Campbell, et al., "A Transient Thermal Model of the Portable Life Support System," Paper No. 94-4622, AIAA Space Programs and Technologies Conference, Huntsville, Alabama, September 27-29, 1994.

24. L. Smith, et al., "Human Thermal Model Evaluation for Portable Life Support System Control Development," Paper No. 94-4621, AIAA Space Programs and Technologies Conference, Huntsville, Alabama, September 27-29, 1994.

25. Buyevich, Y.A. and Webbon, B.W., "Effects of Gravity on Bubble Formation at a Plate Orifice," ASME Fluids Engineering Annual Conference, August 13-18, 1985.

26. Buyevich, Y.A. and Webbon, B.W., "Evolution of Vapour Bubbles at Nucleation Sites in Low Gravity," Euromech Colloquium, Institut fur Stromungsmechanik; Universitat Karlsruhe, 1995.

27. A.C. McCormack and B.W. Webbon, "Testing of Electrically Heated Gloves for Cold Environments," 25th International Conference on Environmental Systems, July 10-13, 1995; SAE Technical Paper # 951547.

28. Y. Buyevich and B. Webbon, "Modeling of Vapor Bubble Growth Under Nucleate Boiling Conditions in Reduced Gravity," National Heat Transfer Conference, Portland, Oregon, March 13-16, 1995.

29 Y.E. Ku, J.E. Carbo, L.D. Montgomery, and B.W. Webbon, "Hemodynamic responses to head and neck cooling," presented at the 66th Annual Scientific Meeting of the Aerospace Medical Association, Anaheim, California, May 7-11, 1995.

30 B.W. Webbon, L.D. Montgomery, and R.K. Callaway, "Biomedical Use of Aerospace Personal Cooling Garments," presented at the 66th Annual Scientific Meeting of the Aerospace Medical Association, Anaheim, California, May 7-11, 1995.

31 Y.E. Ku, L.D. Montgomery, and B.W. Webbon, "Hemodynamic and Thermal Responses to Head and Neck Cooling in Men and Women," presented at the 1995 Annual Meeting of the Consortium of Multiple Sclerosis (MS) Centers, Portland, Oregon, September 14-17, 1995.

32. Y. Buyevich and B. Webbon, "Effects of Zero Gravity on Bubble Formation at a Plate Orifice," presented at the 1995 International Mechanical Engineering Congress & Exposition, San Francisco, California, November 12-17, 1995.

33. Y. Buyevich and B. Webbon, "Evolution of Vapor Bubbles at Nucleation Sites in Low Gravity," Euromech Meeting on Flows with Phase Transitions, Gottingen, Germany, August 5-9, 1995.

34 Y.E. Ku, L.D. Montgomery, C. Lomax, and B.W. Webbon, "Liquid Cooling Garment Technology Transfer: a Biomedical Case Study," presented at the 67th Annual Scientific Meeting of the Aerospace Medical Association, Atlanta, Georgia, May 5-9, 1996.

35. A.B.Campbell, S.S. Nair, J.B. Miles, and B.W. Webbon, "PLSS Transient Thermal Modeling for Control," 26th International Conference on Environmental Systems, SAE Technical Paper #961482, July 8-11, 1996.

36. P. Stone, J.B. Miles, S.S. Nair, and B.W. Webbon, "Tubular Membrane Evaporator Development for the PLSS," 26th International Conference on Environmental Systems, SAE Technical Paper #961486, July 8-11, 1996.

37. L. Smith, S.S. Nair, J.B. Miles, and B.W. Webbon;,"Evaluation of Human Thermal Models for EVA Applications" 26th International Conference on Environmental Systems, SAE Technical Paper #961487, July 8-11, 1996.

38. S.B. Thornton, D. Storm, S.S. Nair, J.B. Miles, and B.W. Webbon, "Requirements and Accuracies in Human Exercise Measurement," 26th International Conference on Environmental Systems, SAE Technical Paper #961532, July 8-11, 1996.

Other Publications and Reports:

1 B.W. Webbon, "Redesign of Turbine Aerodynamic Plane Cascade," Pratt & Whitney Aircraft Internal Report, May 1968.

2 B.W. Webbon, "Summary of the Results of Metal to Gas Temperature Ratio on Turbine Cascade Profile Loss," Pratt & Whitney Aircraft Internal Report, August 1968.

3 B.W. Webbon, *Apollo 13 Radiator Performance Prediction,* Vought Missiles and Space Company D.I.R. No. 350-DIR-28, December 1969.

4 B.W. Webbon, *Review of the PLSS Thermal Model O_2 and H_2O Flow Systems Simulation, EMU Digital Simulator,* Vought Missiles and Space Company D.I.R. No. 00-DIR-254, July 1970.

5 B.W. Webbon, *Cabin Condensation Model Update for MSD-T Thermal Simulators,* Vought Missiles and Space Company D.I.R. No. 350-DIR-35, January 1971.

6 B.W. Webbon, *Concept Evaluation Test of a Fusible/Evaporative Astronaut Heat Sink,* Vought Missiles and Space Company Report No. 00.1452, July 1971.

7 B.W. Webbon, *Advanced Extravehicular Protective Systems Emergency Considerations,* Vought Missiles and Space Company Report No. T141 RP004, September 1971.

8 J. Williams, B.W. Webbon, and R. Copeland, "Advanced Extravehicular Protective Systems Study," Interim Report NASA CR-114321, 1971.

9 J. Williams, B.W. Webbon, and R. Copeland, "Advanced Extravehicular Protective Systems (AEPS) Study," Final Report NASA CR-114382,March 1972.

10 B.W. Webbon, *Prebreathing Data Summary for Use in the VMSC Space Shuttle EVA Equipment Study,* Vought Missiles and Space Company D.I.R. No. T-192-DIR-01,July 1972.

11. B.W. Webbon, et al., *Study of Space Shuttle EVA/IVA Support Requirements, Tasks, Guidelines, and Constraints Definition,* NASA CR-133992, April 1973.

12. H.C. Vykukal and B.W. Webbon, *High Pressure Space Suit Assembly,* NASA TM-73089, December 1975.

13. B.W. Webbon, "A Zero Gravity, Nucleate, Pool Boiling Experiment," AIAA 11th Thermophysics Conference Open Forum, July 1976.

14. B.W. Webbon, et al., "A Comparison of 3 Liquid/Ventilation Cooling Garments During Treadmill Exercise," published in *Proceedings of Aerospace Medical Association 50th Annual Scientific Meeting,* May 1979.

15 R. Rubenstein, J. Kurzweil, and B.W. Webbon, "Preliminary Report: Microprocessor Based Automatic Control of Human Thermal Regulation with a Liquid Cooled Garment," NASA Ames Research Center/San Jose State University Joint Research Interchange No. NCA Z-0R675-904, February 1980.

16 B.W. Webbon, "Survival in Space," *Natural History Magazine,* pp. 50-57, December 1981.

17 C.S. Weaver, B.W. Webbon, and L.D. Montgomery, "Development of a Prototype Automatic Controller for Liquid Cooling Garment Inlet Temperature," final Contract Report NASA Contract NAS 9-16487, September 1982.

18 *Spaceplane Concept Feasibility Examination,* SD-TR-83-45, limited distribution, May 1983.

19 B.W. Webbon, "A Comparison of Manned and Unmanned Orbital Construction and Maintenance," presented at 20th Space Congress, Cocoa Beach, Florida, April 1983.

20. B.W. Webbon, *Space Station Rescue Vehicle Study,* SRI Technical Report, Project No. 5095, February 1983.

21 J.G. Depp, J.D. Malick, and B.W. Webbon, *Military Applications of a Manned Space Platform,* SRI Final Report (Classified), Project No. 5095, January 1983.

22 B.W. Webbon, et al., "Possible Uses of Various Physiological Monitoring Systems in the Scientific/Life Sciences Laboratory of the Space Station," SRI Project No. 86557 Final Report Task 2.0, January 1986.

23 B.W. Webbon, "An Evaluation of Automatic Control System Architecture for EVA Portable Life Support Systems," Lakeview Research Consulting Report, private client, September 1986.

Mrs. MORELLA. Thank you for those success stories and anecdotes, Dr. Webbon.

I'm now pleased to recognize Steve Jacobs. Thank you, Mr. Jacobs.

Let me also comment that we've been joined by Mr. Cannon from Utah.

TESTIMONY OF STEVE JACOBS, EXECUTIVE ASSISTANT TO THE PRESIDENT, NCR CORPORATION, DAYTON, OH

Mr. JACOBS. Thank you, Madam Chairwoman and members of the Subcommittee on Technology, for providing me the opportunity to discuss the business benefits of meeting the needs of people with disabilities through technology transfer.

My name is Steve Jacobs. I'm a Senior Technology Consult with the NCR Corporation. NCR is based in Dayton, Ohio, as is Wright Patterson Air Force Base.

Over the past 4 years, NCR technologists have met on many occasions with Wright Lab scientists and Wright State University engineering students to share our knowledge and visions for the future.

I mention Wright State for an important reason: Like NTSU, based on Dayton, Ohio, Wright State University has one of the largest numbers of students as a percentage of total enrollment of any university in the United States of students with disabilities.

Developing products that are accessible, usable, and useful by people with disabilities brings more benefits to mainstream business than may be obvious to the casual observer. The business work refers to products such as these as universally designed products.

A universally designed product is defined by the business world as a product that is accessible, usable, and useful by people with a wide range of abilities in a wide range of situations. There's not much difference between technologies developed to enable a foot soldier to access computer-based information at night, and technologies that enable people who are blind to access computers.

Worldwide, there are 154 million consumers who are blind and low-visioned. Thanks to speech synthesis, originally developed as an assistive technology, people who are blind and low-visioned, can now read with their ears as can our foot soldiers.

Text-to-speech technologies have other important business implications. For example, there's little difference between a person who is blind and a person who is illiterate from the standpoint of not being able to read.

Worldwide, there are more than 1.1 billion consumers who are illiterate. This can be a real market limiter for companies wishing to market publicly-accessible information systems on a global basis.

Speaking of people who are unable to read, I cannot read. Of course, I can read in English, but not in any other language.

Actually, it would not be inappropriate to consider me a person with a disability from the standpoint of not being able to access and use foreign language-based information systems.

This is an important point in light of the fact that there are nearly half a billion tourists traveling to foreign countries each year. The number of foreign-born citizens living in the United States will grow to more than 29 million people by the year 2000.

Extrapolating these data to a worldwide level brings the number of people who may not be fluent in the language native to the information system they may wish to use to 300 million consumers.

The assistive technology of multilingual speech synthesis has the potential to enable these individuals to benefit from the information age in ways that were never before possible. If one were to add the total number of people represented by each of the consumer groups I just mentioned, including overlap, we're talking about more than two billion consumers.

There's a technology trend that is beginning to disable people. Electronic consumer products are getting smaller.

Among these items are pagers, cell phones, laptop computers, personal digital assistance, palmtop computers and smart phones. From a competitive standpoint, the smaller, the better.

But there's just one problem. The smaller these devices become, the more difficult it is to read their displays and use their controls.

The inability to use these products is further compounded by the fact that the average consumer is getting older. The older we get, the more likely we are to have sight, mobility, hearing, and memory problems.

Speaking about aging, people are living longer. The 60-plus age group will make up 16 percent of the total world population by the year 2030.

There were 16 workers for every retiree in 1950. This ratio is expected to fall to one worker for every two retirees by the year 2010.

Today, worldwide, there are approximately 360 million consumers who are 65 years of age or older. This is not a small consumer group.

Thanks to advanced technologies like synthesized speech, voice recognition, and non-visual Web browsers being pioneered by Productivity Works, our senior citizens do not have to be disenfranchised from our information society.

Not only is the average worker getting older, they are becoming more mobile. It is not easy using cellular telephones and laptop computers while driving a car, unless of course, you don't mind getting in any accidents.

However, help is on the way. New neural digital signal processing technologies are being developed in disability research labs across the country.

These devices and the types of devices being pioneered by companies such as Biocontrol are now being used to develop low-cost, wireless input devices for mainstream use.

Thanks to products such as these, neural signals generated by the brain, eye, and muscles, can now be used to enhance busy environments to control cellular phones, laptop computers, and many other electronic devices.

It is for this purpose that neural signal devices are being developed by Grant McMillan and his colleagues at Wright Labs. Dr. McMillan's work is focused on enabling fighter pilots to adjust noncritical controls in their fighter aircraft without needing to use their hands.

Supporting programs which foster technology and knowledge transfer between rehabilitation engineering centers, federal laboratories and big business have many potential benefits.

In addition to bridging the widening gaps I've just cited, from a business perspective, collaboration among scientists from many disciplines also stands to shorten product development life cycles, reduce costs, and increase quality. These benefits stand to benefit all of us.

It is my hope, Chairwoman Morella and members of the Subcommittee on Technology, that you support collaborations such as these to the greatest extent possible.

I would like to close my testimony with a quote by a young man I worked with 4 years ago. His name is Randy Gilbert. He is a software programmer who just happens to be paralyzed from the neck down.

When asked what he thought about being disabled, he said, "Disabilities only appear in the eyes of the beholder. They disappear through the eyes of the innovator."

Thank you again for providing me this wonderful opportunity to testify before your Subcommittee.

[The prepared statement of Mr. Jacobs follows:]

Testimony of Steven I. Jacobs
Senior Technology Consultant
NCR Corporation
before the
United States House of Representatives
Committee on Science
Subcommittee on Technology
July 15, 1997
2:00 PM
In Room 2318 of the Rayburn Building
on
"Meeting the Needs of People with Disabilities through Federal Technology Transfer."

Thank you Chairwoman Morella, and members of the Subcommittee on Technology, for providing me the opportunity to discuss the business benefits of "Meeting the Needs of People with Disabilities through Federal Technology Transfer."

My name is Steve Jacobs. I am a Senior Technology Consultant with the NCR Corporation. NCR is based in Dayton, Ohio, as is Wright Patterson Air Force Base. Over the past four years, NCR technologists have met, on several occasions, with Wright Lab scientists and Wright State University engineering students to share our knowledge and visions for the future. I mention Wright State for an important reason. Based in Dayton, Ohio, Wright State University has one of the largest number of students with disabilities, as a percentage of total enrollment, of any university in the United States.

Developing products that are accessible, usable and useful by people with disabilities brings more benefits to mainstream business than may be obvious to the casual observer. The business world refers to products such as these as universally designed products. A universally designed product is defined, by the business world, as a product that is accessible, usable and useful by people with a wide range of abilities, in a wide range of situations.

There is not much difference between technologies that have been developed to enable a foot soldier to access computer-based information at night and technologies that enable people who are blind to access and use computers. Worldwide there are 154 million consumers who are blind and low-visioned. Thanks to text-to-speech technologies, originally developed as an assistive technology, people who are blind can now read with their ears as can our foot soldiers.

Text-to-speech technologies have other important business implications. For example, there is little difference between a person who is blind, and a person who is illiterate, from the standpoint of not being able to read. Worldwide there are more than 1.1 billion consumers who are illiterate. This can be a real market-limiter for companies wishing to market public access information systems on a global basis.

Speaking of people who are unable to read, I can't read. Of course I can read English but not any other language. In fact, it wouldn't be inappropriate to consider me a person with a disability from the standpoint of not being able to access and use foreign-language based information systems.

This is an important point in light of the fact that nearly half-a-billion tourists travel to foreign countries each year.

The number of foreign-born citizens, living in the United States is expected to increase to more than 9% of our population by the year 2000. This will represent nearly 30 million consumers. Extrapolating these data to a worldwide level brings the number of people, who may not be fluent in the language native to the information systems being used in the countries in which they live, to 300 million consumers. The "assistive technology" of multilingual text-to-speech synthesis has the potential to enable these individuals to benefit from the information age in ways that were never before possible.

If one were to total the number of people represented by each of the consumer groups I just mentioned, including overlap, we are talking about more than 2 billion consumers.

But it doesn't stop there. Consumer electronic products are beginning to disable people. These technologies include pagers, cell phones, laptops, desktops, personal digital assistants, palmtop computers and smart phones. All of these products are getting smaller. From a competitive standpoint, the smaller the better. There's just one small problem. The smaller these devices become, the more difficult it becomes to read their displays and use their controls. The ability to use *shrinking* devices is compounded by another small problem. The average person using them is getting older. And, unfortunately, along with increased age comes increased disabilities.

People are living longer; The sixty-plus age group will make up 16% of the total world population by 2030; There were sixteen workers for every retiree in 1950. This ration is expected to fall to just 2:1 by 2010. Today, worldwide, there are an estimated 360 million consumers 65 years of age and older. Not a small consumer group. Were it not for assistive technologies, being brought into the mainstream, many of our senior citizens would be disenfranchised from our information society.

Not only is the average consumer getting older, they are becoming more mobile. It's tough using cell phones and computers while driving a car unless, of course, you like hurting yourself.

New neural digital signal processing technologies are being developed in disability research labs across the country. These devices, and the types being pioneered by companies such as BioControl, are now being used to develop low cost wireless input devices for mainstream business use. Using neural signals generated by the brain, eye, or muscles to control computer functions is now possible.

These types of input devices work quite well in mobile, hands-busy environments. For that reason, it's no small coincidence that these technologies are also being pioneered by scientists such as Grant McMillan and his colleagues at the Alternative Control Technology Laboratory of Wright-Patterson Air Force Base in Dayton to enable fighter pilots to adjust non-critical controls in their fighter planes, without needing to use of their hands or voices.

I am convinced that strengthening the ability to meet the needs of people with disabilities through Federal technology transfer would bring with it many benefits. I believe supporting programs which proactively encourage collaboration between rehabilitation engineering centers and Federal Laboratories could shorten product development life-cycles, reduce costs and increase the quality of products currently under development.

It is my hope, Chairwoman Morella, and members of the Subcommittee on Technology, that you support efforts such as these to the greatest extent possible.

I would like to close my testimony with a quote from a young man I worked with four years ago. His name is Randy Gilbert. He's a software programmer who just happens to be quadriplegic.

When asked what his thoughts were on being disabled he said, "Disabilities only appear in the eyes of the beholder; they disappear through the eyes of the innovator."

Thanks, again, for your time and for this wonderful opportunity to address your subcommittee.

Steve Jacobs
NCR Corporation

Steve Jacobs is a Senior Technology Consultant with the NCR Corporation. Previous to his current assignment Jacobs' spent six years managing a variety of NCR technical support, education and software development organizations. Jacobs also spent two years as a hardware product manager; four years as a commercial industry marketing manager and five years in system sales.

In addition to his technology consulting role, Jacobs has assumed a leadership role in promoting the business benefits of universal design and information accessibility for people with disabilities. Jacobs is President of NCR IDEAL, a Business Resource Group focused on meeting the needs of NCR employees and customers with disabilities. Jacobs' other activities, in this area, have included:

Serving (3 years) as an active participant in AT&T's Consumer Advisory Panel on Disability Issues.

Serving as a member of the Telecommunications Access Advisory Committee (TAAC). TAAC was convened, in support of the FCC, to develop guidelines for access to telecommunications equipment and customer premises equipment;
http://www.access-board.gov/telecom/telecom.htm
http://trace.wisc.edu/text/telecomm/august96/taac_I.html
http://trace.wisc.edu/text/telecomm/sept96/schroedr.txt

Serving as a member of the White House's Web Access Initiative (WAI) planning committee. The WAI program was created to develop an International Program Office to coordinate World wide Web-related activities; http://www.w3.org/WAI/References/access-brief
http://w3c1.inria.fr/Press/Clinton.html
http://www.w3.org/TandS/Public/NSF-Accessibility-Proposal.html,
http://www.w3.org/WAI/References/9703/NSF.html

Serving (current) as a member of WAI, to enhance the accessibility of web formats and protocols (HTML, XML. CSS, HTTP).
http://www12.w3.org/WAI/Overview.html

Partnering in the development of the unified Web Development Guidelines;
http://www.trace.wisc.edu/HTMLgide/htmlfull.html

Managing the implementation of a, universally designed, web-based resource center in support of the 53rd Presidential inauguration;
http://www.prodworks.com/ncr
http://www.prodworks.com/ncr/journ6.htm,

Managing the implementation of a, universally designed, web-based resource center for the United Nations' International Leadership Forum. This Forum, Chaired by First Lady Hillary Rodham Clinton, was the follow-up meeting to the United Nation's Fourth World Conference on Women held in Beijing, China, in 1995; http://www.prodworks.com/ilf
http://www.prodworks.com/ilf/credits.htm
http://www.prodworks.com/ilf/nsf.htm

Serving (current) as a member of the National Information Standards Organization (NISO) digital talking book standards development committee. This committee is defining the functional requirements for the next generation digital media. This technology will have far reaching effects on the way audio, in digital format, can be searched and played in an accessible manner over the Internet;
http://www.loc.gov/nls/news-rel-97-02.html

Participating in the President's Electronic Commerce initiative by demonstrating web technologies which have been developed for universal access;
http://www.whitehouse.gov/WH/New/Commerce/,
http://www3.zdnet.com/zdnn/content/inwk/0422/inwk0009.html,
http://www.ncr.com/press_release/pr070197.html,
http://www.services.ncr.com/industries/telecom/commarkt/articles/cibu/97jun/0630b.htm

Serving (current) as a consultant to the United Nations' working group on Informatics and Communication on ways the United Nations can maximize the accessibility of their 21st century computerization program.

Addressing the Federal CIO Consortium on Web Accessibility and E-Commerce Web-based technology.
http://www.itpolicy.gsa.gov/mkm/irmco/notice97.htm

Managing the implementation of a, universally designed, web-based resource center in support of the Department of Labor's One-Stop Office of the Future conference.
http://www.jettcon.org/highres/

<u>Memo:</u>

Date: July 14, 1997

From: Steve Jacobs

To: The Honorable Constance Morella, Chairwoman, Subcommittee on Technology

Subject: Receiving Grant Monies

To the best of my knowledge NCR Corporation receives no grant monies from the Federal Government.

Steve Jacobs
NCR Corporation

Mrs. MORELLA. Thank you very much for your excellent testimony, Mr. Jacobs.

I'm now going to turn to, if I may, Mr. Hershberger, than Mr. Lahoud will be the final panelist.

TESTIMONY OF DAVID H. HERSHBERGER, VICE PRESIDENT OF PRODUCT DEVELOPMENT, PRENTKE ROMICH COMPANY, WOOSTER, OH

Mr. HERSHBERGER. Chairwoman Morella and members of the Subcommittee on Technology, I want to thank you for inviting me to testify at this hearing on meeting the needs of people with disabilities through federal technology transfer.

My name is Dave Hershberger. I'm Vice President of Product Development at the Prentke Romich Company.

Prentke Romich was founded in 1966 for the sole purpose of providing technology for people with disabilities. This has remained the company's only activity throughout its history.

Today, this Ohio-based company employs 150 people throughout the United States, and supplies assistive devices throughout the United States and also exports an increasing number of products.

Prentke Romich's primary products are speech generation devices for people who cannot speak. These augmentative communication devices enable many people who were previously unable to communicate basic needs to now be able to express their thoughts in written and spoken form, thus giving them greater opportunities to participate in activities such as going to school or becoming employed.

Technological developments have had a profound role in assistive devices for people with disabilities. Many of the products available today might have been inconceivable with the technology of even 20 or 30 years ago.

Rehabilitation Engineering Research Centers, or RERCs, which are funded through NIDRR, have contributed to the current level of assistive technologies.

The first microprocessor-based augmentative communications system was developed through a RERC, as was the first optical head pointer and the first female voice to be used in augmentative communications systems.

Other advancements come from developments within the assistive technology industry. Like many medium-sized technology companies, Prentke Romich Company has always had a strong research and development component.

Most of our research is self-funded and we've also received some funding from federal SBIR programs.

Because of the relatively small size of our market, we must take advantage of technological developments occurring elsewhere and focus our activities in areas where other industries are less likely to spend their own research dollars.

Despite all of the current efforts applied to research in assistive technologies, we recognize that there is much more that can be done to address the needs of people with cognitive and physical disabilities.

The more that is understood about technology and the needs of people with disabilities, the greater these opportunities appear.

I am encouraged that the role of federal laboratories relative to assistive technology is being discussed.

Based on our past experiences, additional research and development applied to assistive technology will lead to making the lives of individuals with disabilities more fulfilling and productive.

In examples that have already been cited, develops of the RERCs help seed and advance the augmentative communication device industry.

Additional technology from the federal laboratories may have an equal benefit in numerous assistive technology areas.

Transferring these developments from the laboratories into assistive technology should cost substantially less than duplicating the research at the RERCs or at the companies.

Meanwhile, the expertise of the assistive technology companies could be tapped to make this technology more suitable and more affordable for consumers, and also to set up means of distribution, training and support for end users.

Second, greater information transfer will be beneficial. information from RERCs and others within the disabilities field will create greater awareness at the federal laboratories of the technological hurdles faced by people with disabilities.

Likewise, greater knowledge and developments at the federal laboratories will help those of us who develop assistive technology.

Much of this information is aided by the worldwide web, and is already underway.

As the ADA and recent reauthorization of the IDEA confirm, we are privileged to live in a country that acknowledges that people with disabilities have the same rights as all Americans.

For the past 15 years, I've had the opportunity to work in the field of assistive technology, and have seen many individuals live more fulfilling lives, despite their disabilities through the use of this technology.

I commend you for exploring possibilities of enriching the lives of these individuals even further. Thank you.

[The prepared statement of Mr. Hershberger follows:]

Written Testimony for

Meeting the Needs of People with Disabilities through Federal Technology Transfer

to:
United States House of Representatives
Committee on Science
Subcommittee on Technology

2:00 p.m.
July 15, 1997

by:
David H. Hershberger
Vice President of Product Development
Prentke Romich Company
1022 Heyl Rd.
Wooster, OH 44691
Phone: 1-800-262-1984
Fax: 330-262-5586
email: dhh@prentrom.com

Introduction:

Chairwoman Morella and members of the Subcommittee on Technology, I want to thank you for inviting me to testify at this hearing on "Meeting the Needs of People with Disabilities through Federal Technology Transfer." My name is Dave Hershberger and I am Vice President of Product Development at the Prentke Romich Company.

Prentke Romich Company and Augmentative Communication:

Prentke Romich Company was founded in 1966 for the sole purpose of providing technology for people with disabilities. This has remained the company's only activity throughout its history. Today, this Ohio-based company employs 150 people throughout the United States. Prentke Romich Company supplies assistive devices throughout the United States and also exports an increasing number of products.

Prentke Romich Company's primary products are speech generation devices for people who cannot speak. Using microcomputer technology, a language organization system called Minspeak and thirty years of experience, we develop and manufacture devices which allow people to generate speech by pressing keys on a keyboard, moving a joystick, pointing their head or using virtually any controlled muscle movement. These devices, often referred to as Augmentative Communication Devices, enable many people who were previously able to communicate only basic needs to their attendants to have much greater communication opportunities. They can join in conversations, write letters, take notes, give speeches, use the telephone and more recently, participate in electronic communication through the Internet. Providing a means of communication opens doors for these individuals. Many now have the opportunity to attend school and/or to become employed.

Advancements in Technology:

Technological developments have had a profound role in assistive devices for people with disabilities. Many of the products available today would have been inconceivable with the technology of twenty years ago. Some of these innovations have been developed by assistive technology companies, some by Rehabilitation Engineering Research Centers and many others by developers in the consumer market.

The computer revolution has resulted in a lightning pace in the advancements of electronics and technology. This revolution has had a profound impact on people with disabilities. First, although not universally the case, many of the new products are easier to use, or at least to modify for use, by individuals with disabilities. Consider the remote controls which allow us to control the TV set from our armchairs. These conveniences can also be used by people who are unable walk across the room to change channels. A second benefit of accelerated advancement in more sophisticated consumer devices is the accompanying advancement in the components required to make the devices. The components required to produce faster and better computers are also used to make more sophisticated wheelchairs controls or augmentative communication aids.

Rehabilitation Engineering Research Centers (RERCs) which are funded through NIDRR have also contributed to the level of assistive technology available today. The first microprocessor-based augmentative communication aid was developed at the Trace Research and Development Center at The University of Wisconsin. The first optical headpointer and the first female voice to be used in augmentative communication systems were also developed at RERCs.

Other advancements come from the developments within the assistive technology companies. Like many medium sized technology companies, Prentke Romich Company has always had a strong Research and Development component. As a commercial company, most research activities are funded through moneys generated through product sales. We have also received funding through the Small Business Innovation Research Program. Because of the relatively small size of our market, we must be judicious in how we spend our research and development dollars. Specifically, we must take advantage of technological developments occurring elsewhere and focus our own efforts toward adapting the technology so that it can be used by people with disabilities. For example, rather than developing the next generation of speech synthesizers, we concentrate on implementing the latest synthesizers into equipment that can be operated by individuals with severe disabilities. Also, assistive technology companies must concentrate their efforts in areas where other industries are less likely to spend their own research dollars. Research projects in which Prentke Romich has participated in recent years include:
- specialized scanning techniques for faster speech generation
- hands-free computer operation
- augmentative communication devices that can be worn by an individual
- visual language representation

Opportunities:

Despite all of the current efforts applied to researching assistive technology, we believe that there is much more that can be done to address the needs of people with physical and cognitive disabilities. The more that is understood about technology and the needs of people with disabilities, the greater the opportunities appear. As an example, Prentke Romich Company is working jointly with a team at the Applied Science and Engineering Laboratories at the University of Delaware (also an RERC) to implement artificial intelligence into communication devices. The goal of the project is to take the telegraphic and incomplete speech from children with low cognitive skills and generate complete, grammatically correct sentences that can be understood by a greater number of people.

I am encouraged that the role of the National Laboratories relative to assistive technology is being discussed. Based on past experiences, additional research and development applied to assistive technology will lead to making the lives of individuals with disabilities more fulfilling and productive. From the perspective of an assistive technology company, I can see several opportunities for cooperation between the National Laboratories and the RERCs in making technology available to the people who would receive the greatest benefit from it.

In the examples that have already been cited, developments at RERCs helped seed and advance the Augmentative Communication device industry. Additional technology from the Federal Laboratories may have an equal benefit in numerous assistive technology areas. Transferring these developments from the Laboratories into assistive technology should cost substantially less than duplicating the research in the assistive technology companies or RERCs. Meanwhile the expertise of the assistive technology companies can be tapped to make the technology more suitable and affordable for consumers and to set up means of distribution, training and support for the consumers.

Secondly, greater information transfer would be beneficial. Information from RERCs, assistive technology companies and others within the disabilities field could create greater awareness at Federal Laboratories of the technological hurdles faced by people with disabilities. Likewise, greater knowledge of developments at Federal Laboratories would help those of us who develop assistive technology. Much of this information exchange is aided by the World Wide Web and is already under way.

Conclusions:

As the ADA and recent reauthorization of IDEA confirm, we are privileged to live in a country that acknowledges that people with disabilities have the same rights as all Americans. For the past fifteen years I have had the opportunity to work in the field of assistive technology and have seen many individuals live more fulfilling lives despite their disabilities with the use of this assistive technology. I commend you for exploring possibilities of enriching the lives of these individuals even further.

Federal Grants for Development of Assistive Technology in which Prentke Romich Company has participated.

From:
National Institute of Health
National Institute on Deafness and other Communication Disorders
Small Business Innovation Research Program

A Communication Aid Enhanced with Semantic Parsing
Reference: 1 R41 DC02338-01A1
Project Period 9/30/95 - 9/29/96
Funds received (or to be received): $79,212

Innovation in Augmentative Communication Devices
Reference: 2 R44 DC02211-02
Project Period 1/1/94 - 5/31/97
Funds received (or to be received) Phase 1: $75,000
Funds received (or to be received) Phase 2: $750,000

Development and Testing of Portable Communication Aids
Reference: 1 R43 DC03251-01
Project Period 5/1/97-10/31/97
Funds received (or to be received): $82,986

Optimization of Speech Synthesis Software for Vocal Communication Aids
Reference: 5 R44 MH52357-02
Funds received (or to be received): 0

From:
United States Department of Education

Assessing the Utility of the Child-Oriented Readily Expanding Words Strategy
Reference: 93-025
Project Period 9/1/93 - 2/28/94
Funds received (or to be received): $40,000

<u>David H. Hershberger</u>

David H. Hershberger is Vice President of Product Development for Prentke Romich Company. Mr. Hershberger has been with Prentke Romich Company since 1981 and has held several positions in manufacturing and product development. In his current position, he leads the company's engineering, research and product development efforts. Mr. Hershberger earned a B. S. degree in Electrical Engineering from The University of Akron. He has published papers in various conference proceedings and archival journals and has been a presenter at numerous professional conferences.

Mrs. MORELLA. Thank you very much, Mr. Hershberger.

Now, as I recognize Joseph Lahoud, who is the President of LC Technologies, I will say that the reason I put you at the end, too, is that Congressman Tom Davis was going to be here for your presentation.

We'll make sure that he sees it, since he represents you in Fairfax, Virginia. So, Mr. Lahoud.

TESTIMONY OF JOE LAHOUD, PRESIDENT, LC TECHNOLOGIES, FAIRFAX, VA

Mr. LAHOUD. Thank you, Madam Chairwoman and members of the Subcommittee.

I appreciate the opportunity to be here to talk about assistive technology and the role that small business plays in this field. My company, LC Technologies in Fairfax, I believe, is a typical example of a small company in the United States that is involved in assistive technology, the research and development associated with assistive technology, and the search for the really enormous resources that are required to bring new products and new assistive technologies to the marketplace.

We are a very small company. We range from 8 to 12 people, depending upon what time of the year you're talking to us, and what's going on in the company.

The company is 11 years old. I'm not sure if 11 years ago I would have been able to stick it out if someone had told me that it was going to take 11 years to get to the stage where we are now.

But we have stuck it out. To me, it's the most rewarding thing that anyone could ever do, is to get to the point where we are today.

Al lot of it was serendipity. In these last couple of years, we've been able to take advantage of the experience that we've had over the years of developing and attempting to commercialize this technology, to understand better what's happening in our country, what's happening in the public sector, the Federal Government, state governments, and how the private sector views assistive technology and helping people with disabilities.

We did not start out to develop an assistive technology. We were a research and development company doing a variety of research projects in a fancy area of technology called signal processing and image processing.

We, the we being one other individual and myself with technology backgrounds that had worked in many, many different fields, and while working casually one day or in a casual conversation with a friend from a university in Virginia we learned about this technology called automatic eye tracking.

We had no idea that it ever existed or that there was such a thing. But we found out that way back in the 1960's—and I think this is an early example of technology transfer, a technology that was actually developed as a result of the Department of Defense support and motivation way back in the 1960's.

The Air Force was interested in seeing if there are alternate means of allowing pilots and other workers to interface with their equipment while their hands are busy and their voices are busy or the environment does not allow them to use their hands or their voice.

And it was an obvious thing to look at the eyes. Is there a way that we can monitor the eye movements of people, and from what we learn from those eye movements, give people an opportunity to control things that otherwise they would not be able to control?

And it was that work sponsored by the Air Force at Wright Patterson Air Force Base that resulted in the original invention of the method that is the most commonly used method today to determine exactly what people are looking at when your eyes are being viewed by a video camera.

There were some significant technological obstacles that no one was able to overcome back in the 1960's and 1970's, which as kept this technology from ever getting into the real world. Other people tried and made additional developments.

When we came along in 1986 and 1987, we were able to build, begin with what other people had done, and were able to make some breakthroughs in a relatively short period of time, which advanced the technology beyond where it had been before.

And when we knew early on what could be done with this technology, particularly the accuracy, how accurately you can monitor someone's eye movement and tell what they're looking at, it was very obvious that the greatest beneficiaries of this technology would be people with severe physical disabilities who are otherwise not able to use their hands or even to use their voice.

The community of quadriplegics whose eyes are normal, even only one eye is all that's necessary, and their minds are still intact, otherwise, they're totally locked in and unable to be productive, to be independent in any way.

This technology offered an opportunity for that group to become completely productive with the computer and to do things that they never otherwise would be able to do.

So, with the accuracy that we developed, we were able to display an entire keyboard on a computer screen at one time, and when you look at those individual keys, anywhere from 70 to 100 keys, the system is accurate enough to recognize what you're looking at.

And by just keeping your gaze inside a key for a fraction of a second, that key is executed as if you did it with your finger on a keyboard.

As a small company, we developed the technology to that point in the first couple of years with our own resources. It didn't take long to run out of those resources, and knew that now we were going to be dependent on trying to get some products out into the marketplace, and we were successful in doing that.

But it wasn't enough to keep the company going. It certainly was not enough to keep the company alive, and at the same time provide the resources to make the other developments that would be necessary to get the technology into that point in the future where a much larger population of people would be able to use it.

So, it was during this point in time in the late 1980's and early 1990's that we began to communicate enough with other people in the country to see the changing attitudes on the part of helping people with disabilities, both in the government and in the private sector.

And we saw opportunities within the Federal Government to get some of the resources, to collaborate with others to get the resources that are necessary to develop this technology into the future.

Today, the two groups that are well represented here and are responsible for this hearing; the NIDRR group headed by Dr. Seelman, and the FLC, Mr. Taylor as the Washington representative to the FLC, are two very, very key people in establishing this very expanding and more open attitude on the part of the Federal Government to use its resources in the federal laboratories and elsewhere to further the benefits for people with disabilities.

We have participated as a small company in the SBIR program, which is a tremendously valuable program for small business, and many of the opportunities in SBIR are helpful for people with disabilities, but that needs to be expanded.

The STTR program is also important. It's a relatively new program, but it also offers opportunities for small business to get some of the resources that we need.

But we knew that even if we had all the resources that we needed in the way of funding, you still need other resources, other expertise, other things that people have done in the past that we would like to take advantage of. So, collaborating and particularly collaborating somehow through the federal laboratory system, would be the most valuable and important thing for us to do.

And we've made some progress there. In my written testimony, I describe the very good working relationship we have with one of the RERCs funded by NIDRR, the University of Delaware, where we're working hand-in-hand with them to develop eye tracking technology and to make it useful for people with disabilities.

I think it's particularly important in that program to realize that the relationship that exists between LC technologies and the University of Delaware in that program, is not a "we've developed this, now you take it and commercialize it;" it's an ongoing development effort where the University of Delaware stays very close to the community of people with disabilities as we work with them to make sure that the ultimate product and products are useful for the people that are ultimately going to use them.

Through all of the networking that's being done, we became aware of programs, particularly within NASA and the Department of Defense, and it took a few months but we established a technology cooperation agreement with the Jet Propulsion Laboratory which began several months ago.

Right now, it looks like that perhaps is going to be the most significant collaboration that we've been able to put together because of the skills that exist at the Jet Propulsion Lab, and because of some of the background work that they have already done.

One of the most key components of this technology is the camera. Right now, we use in our products—and for those people that have seen it—and I really appreciate Mrs. Morella coming by and looking. In fact, you sat down and ran the system for a few minutes.

You see that camera, you see the computer, you see all this equipment sitting on the table. It's all off-the-shelf equipment. We buy all that stuff and it's more expensive than this system should be. It's bigger than it ought to be, it's not as portable as it should be.

The camera is the most key component to this thing. A camera technology developed by NASA at the Jet Propulsion Lab, specifically for space research or for space applications, and has not been yet completely developed, also is an ideal camera technology for the eye gaze technology, for the automatic eye tracking.

If there was ever an example of a wise collaboration between private industry and the government, it's the JPL example where the camera technology can be further expanded, not only to serve the space needs of NASA, but also to serve the needs of this eye tracking technology for the benefit of people with disabilities.

Other applications: This is not only a technology for people with disabilities; it's a technology that has enormous long-range potential in many other areas of need, and, as it turns out, needs that are helpful to society.

In the area of learning disabilities, it's a tremendous technology to monitor how people read and to tell the difference between good readers and poor readers, and to come up with solutions, potential solutions for helping people to read better.

It's a technology that has the potential to monitor eye movement in a very unintrusive way, and tell the difference between how you move your eyes when you're well-rested, and how you move your eyes when you're tired.

It's a future component of systems that are going to predict fatigue and drowsiness in the workplace, in vehicles, trucks and planes and so forth. It has enormous potential in that area.

The Department of Transportation is interested in that part of it, and so are some of the other agencies.

If I have one particular message or request to bring up here today, it's the importance of this collaboration between the small companies and the federal research laboratory establishment, but also the importance of communication among all the federal laboratories to understand who has what problems that they're trying to solve?

What kinds of resources are being used to solve those problems today? How can we make those resources more effective so that the minimum resources that we have can be put together in such a way to maximize the benefits that come from them?

So it's clear that the Department of Transportation, the Department of Defense, NASA, the Department of Energy, the Nuclear Regulatory Commission, have a strong interest in this technology called automatic eye tracking.

All of the laboratory resources that are involved in all of those different agencies and others, should be communicated, should be understood. How can we put these resources together so that we're not duplicating each other's efforts, and we can get on with the job?

Again, I really appreciate the opportunity to be here to make this presentation, and I hope to work together in the future in any way that I can.

[The prepared statement of Mr. Lahoud follows:]

Committee on Science
Technology Subcommittee
United States House of Representatives

Assistive Technologies Hearing
"Meeting the Needs of People with Disabilities
Through Federal Technology Transfer

July 15, 1997
2:00 - 4:00 PM

**Eyegaze -
An Assistive Technology,
A Multiple Use Technology, and
an Example of Technology Transfer**

**Joseph A. Lahoud, President
LC Technologies, Inc.
Fairfax, Virginia**

I appreciate the opportunity to appear here today to discuss the Eyegaze technology, which is a working model of a collaborative arrangement between a small company and the federal laboratory system.

LC Technologies is a small company in Fairfax, Virginia specializing in the development and commercialization of a unique technology involving the automatic tracking of the movement of the human eye, which we call Eyegaze. By combining a video camera with a computer, the location of a person's gazepoint on a computer screen can be accurately determined. As a fundamental human-computer or man-machine interaction technology, there are many applications of Eyegaze that have both short-range and long-range commercial potential. One of the most important applications, and one which is available on the market today, is a computer/communication system for people with physical disabilities who cannot use traditional keyboards or other computer input devices that require hand and finger movement. With Eyegaze, these people can now be totally productive on a computer system by merely moving their eyes.

There is a long and complex history associated with the development of automatic eyetracking, going back more than thirty years. The initial pioneering work was actually sponsored by the Department of Defense. In the early 1960's, the Air Force was interested in providing alternate means for pilots to interface with their complex equipment. There were some early successes in the laboratory, but a variety of technological obstacles prevented the transfer of automatic eyetracking to the real world.

Building on the work done by the Air Force and by others, and using only our own resources, LC Technologies, starting in the late 1980's, made some important technological breakthroughs. These breakthroughs enabled us to develop and begin marketing the Eyegaze Computer/Communication System for people with disabilities. To date we have delivered more than one hundred of these devices across the country and overseas.

LC Technologies' goals are to continue the development of the Eyegaze technology, to improve and enhance it for its many future applications, but to make certain at all times that the needs of the community of disabled people are met with the best products at the most economical prices. It is critically important that future Eyegaze products be more lightweight, more miniaturized and portable, and more versatile than they are now, and that they get reduced to reasonable price levels.

The principal strategy that LC Technologies is employing at the present time to acquire the resources we need to continue our Eyegaze development, is partnering with other organizations. Initiatives and resources of the Federal Government represent some of the opportunities we have to obtain critical funding and technology expertise. With regard to the use of Eyegaze as an assistive technology for people with disabilities, initiatives of the Federal Government are essentially the only opportunities we have to fund our research and development.

The Small Business Innovation Research Program, the Small Business Technology Transfer Research Program, and a wide variety of programs involving collaborative arrangements between small companies and Federal Government Research Laboratories, are examples of opportunities for LC Technologies. Unfortunately, only a small percentage of the funds available through the above programs address the R&D needs of the community of people with disabilities.

A very important partnering arrangement we have formed during the past two years is with the University of Delaware's Rehabilitation Engineering Research Center (RERC). As one of several RERC's supported by the National Institute on Disability and Rehabilitation Research (NIDRR), and the one designated for Augmentative and Alternative Communication research, we have been receiving valuable assistance and guidance on matters related to assistive technology, specifically our Eyegaze technology, and on the coordination of Eyegaze commercialization activities with the needs of consumers inner marketplace.

It is an extremely unique situation and one that is immensely beneficial to persons with disabilities, for our company to have the opportunity to form a partnership with the RERC at the University of Delaware. This partnership is particularly effective and has enormous long-range potential, because it is an on-going working relationship, rather than a situation where results are handed off after the work is done. We believe the LC Technologies/University of Delaware RERC partnership is just one of many excellent examples of industry/university/government partnerships that demonstrate how maximum returns can be achieved with minimum resources, for the benefit of people with disabilities.

We have heard elsewhere here today that the National Aeronautics and Space Administration has a major commitment to explore ways to expand the use of its space-related research results to other applications. This commitment includes forming collaborative arrangements with the private sector, with a particular interest in the assistive technology industry. LC Technologies recently entered into a Technology Cooperation Agreement with the NASA/Jet Propulsion Laboratory, which will provide us with some key optical, electronic and manufacturing capabilities necessary for the design and fabrication of miniaturized, versatile and low-cost Eyegaze devices.

The entire Federal R&D establishment has a major commitment to a long-range program designed to predict the onset of fatigue, drowsiness and lack of attention in drivers, pilots, air traffic controllers and others involved in critical activities where there is a major concern about safety and human error. The FAA, as a strong leader in this program, is particularly interested in the Eyegaze technology and its potential to play a major role in future fatigue prediction systems.

LC Technologies received a Small Business Innovative Research contract from the Naval Research Laboratory to devise a solution to an ambient light problem that currently constrains automatic eyetracking systems. We demonstrated a workable solution to this problem under our Phase I contract, and we are looking forward to a continuation of this effort.

At this point it is important for me to repeat and to emphasize my earlier words about LC Technologies' commitment, which is to make certain at all times that the needs of the community of disabled people are met with the best products at the most economical prices. All of our collaborative efforts with others, regardless of the specific applications being explored are designed, first and foremost to make the Eyegaze technology more available and more useful for people with disabilities.

As a representative of the emerging assistive technology industry in the United States, I strongly encourage a continuation of the federal government's commitment to devote as many federal research laboratory resources as possible to collaborative efforts with companies such as mine.

Again, I appreciate the opportunity to speak here today, I applaud the work of this Subcommittee, and I look forward to working with you in the future.

LC Technologies, Inc.

9455 Silver King Court * Fairfax, Virginia 22031 * (703) 385-7133 * FAX: (703) 385-7137

Committee on Science
Technology Subcommittee
United States House of Representatives

Assistive Technologies Hearing
"Meeting the Needs of People with Disabilities
Through Federal Technology Transfer

DISCLOSURE STATEMENT FOR LC TECHNOLOGIES

LC Technologies and the University of Delaware's Rehabilitation Engineering Research Center (RERC) presently have a working relationship, but we have no formal technology transfer agreement. LCT has received no funds (other than the reimbursement for purchased equipment) under any grant or contract from RERC.

LC Technologies (LCT) has signed a Memorandum of Agreement with NASA's Jet Propulsion Laboratory (JPL), dated December 3, 1996. JPL and LCT have both performed technical work under the agreement, however LCT has received no funds (other than travel and equipment purchase reimbursement) under any grant or contract from JPL.

Joseph A. Lahoud, President

Date

CURRICULUM VITAE

JOSEPH A. LAHOUD
President of LC Technologies, Inc.

Mr. Lahoud has BS and MS degrees from Rensselaer Polytechnic Institute in Troy, New York. As a founder and President of LC Technologies, Inc. in Fairfax, Virginia, he has principal responsibility for the overall management of the Company's high technology research, development and commercialization efforts. The commercial activity of LC Technologies is the Eyegaze System, which is receiving worldwide attention as a major advancement in the field of man-machine interaction. The first application of the Eyegaze technology developed by the Company is an eye operated computer and communication system for people with disabilities. This device has received numerous awards, including the Computerworld-Smithsonian first prize for excellence in medical technology. Mr. Lahoud has experience in both large and small technology-based companies, and is active in local and national small business activities. He is also actively involved in community programs designed to serve the assistive technology needs of people with disabilities.

Mrs. MORELLA. Thank you. That's a great statement, Mr. Lahoud. I did very much enjoy using Eye Gaze to know that I can work a computer by just simply looking at various spots on the computer.

The automatic eye tracking just sounds terrific. I understand that there will be—60 Minutes, I think, is looking into doing some case study?

Mr. LAHOUD. There is a gentleman, a journalist in Rhode Island with Lou Gehrig's Disease, who is using this system to write a weekly newspaper column, and he's going to be featured on the 20/20 television program. It's supposed to be shown this coming Friday evening.

Mrs. MORELLA. Very good. Very good.

I want to thank all of you. I thought you did a terrific job. I'm just going to ask a couple of questions, so I can give my colleagues an opportunity, too, and maybe even allow like 15 minutes at the end for people, for my colleagues to go down and see the exhibition that will be closing at 4 p.m.

This topic today is one that's always been of great interest to this Subcommittee, assistive technologies.

Beginning with the landmark 1980 Stevenson-Widener Technology Innovation Act, other legislation between that, and the bill that I introduced in the last session that became law, the National Technology Transfer and Advancement Act, we have been moving forward.

So I wanted to ask—I guess I'll start with Mr. Brand, and then I'm going to ask all of you the very same question.

Where do you believe the direction is that we in Congress should go in our Nation's technology transfer policy? I'm wondering about the perspective from government and from industry about your thoughts about technology transfer from the federal laboratories.

When I say where do we want to go, what direction, I also mean what are the impediments, the barriers, the problems that you think are within our jurisdiction, and what would you suggest that we do?

All right, so I guess I could start with you, Mr. Brand.

Mr. BRAND. Well, Madam Chairwoman, I think that due to your foresight and this Committee's foresight, from the federal community, there are not any legislative impediments to transferring technology from the federal laboratories to industry.

What I see as the impediments, especially in this arena, is lack of awareness. Both in the laboratory from our researchers and engineers, and also in the assistive technology community, we don't always understand the needs from the federal community laboratories, what those needs are.

And we don't see the dual application in many of these technologies. So what we're trying to do is to make more people aware in the laboratories of the need for the assistive technologies and exactly what those are.

We can learn that from our industry colleagues and from the NIDRR laboratories that have been working in this. There hasn't been a relationship between most of those federal laboratories and the NIDRR laboratories in the past.

I think it's a communication thing. As my good colleague from NASA mentioned, we need to know what the assistive technology needs are that are out there. I think it's a matter of awareness.

From your viewpoint, I think it's a matter of stressing the importance to the agencies, the need for assistive technologies within their framework, and the resources that are available for this arena are available to do that, exactly.

So, it's a matter of getting the word out about what the need is and how we can help in doing that.

Mrs. MORELLA. You might think about what it is that we can do make sure that there is the appropriate awareness of what the need is, the fact that it can be fulfilled in our laboratories, and that bridge that brings them together.

Let's go through the panel. Dr. Seelman?

Ms. SEELMAN. I think that there needs to be on both sides, incentives to communicate. I do think there's a potential network of communication there, the locators within the laboratories.

I think that there probably needs to be incentives to exchange actual people and expertise; that is, both the engineers from, for example, NIDRR, or from the industries, and the national laboratories, but also people with disabilities who really need to visit.

Engineers really, when they meet somebody who says I have a problem, I've always found that engineers then like to solve it.

I also think that there needs to be incentives for a pilot project or two. So, with that, I think that incentives will do it.

Mrs. MORELLA. Dr. Webbon?

Mr. WEBBON. Thank you. This was mentioned by some of the other speakers. And it's communication, who is doing what and where.

That's often very hard for companies or individuals or nonprofit organizations who want to access the what they think the Federal Government may have, and they're not sure which way to go.

NASA is in the process of doing basically a technology inventory to really get a nice clean, clear list in one place, of all of the technical activities down in some depth.

Things like that, perhaps on a more government-wide basis would be useful, some sort of a technology inventory. If people could go and they could trace things down to the source, I think it would be very useful.

The idea of transferring personnel is also something we've actually done with the Veterans Administration, among others. But there are some barriers in government agencies, trying to work together.

Transferring funds, for example, between agencies, I've had problems with that. There are major bureaucratic obstacles.

You get money in one fiscal year, but by the time you get it transferred to the other agency that you need to collaborate with, it's the next fiscal year and there are funding problems.

So barriers to spending money and moving money between agencies as part of a collaboration, doing something about those barriers would be useful.

Thank you.

Mrs. MORELLA. Very interesting comment. Doesn't NTIS come out with an inventory also, a technology inventory that maybe should be distributed?

Mr. BRAND. May I add to this? Just this last Thursday and Friday, we had tech transfer conferences in the area where we've looked at the major databases.

There is the DTIC, which is a large inventory database of technologies within the DOD and the DTIN, which is a system within the Department of Energy, and certainly within our FLC, we have a locator system of data inventory.

There are some private databases in addition to the NASA inventory. One of the things under our six focus areas to optimize the diverse resources is to take a look at all of those federal databases and compare those technologies and come up with a common list of technologies that are available in the laboratories.

Still not being able to recognize the need for them in the assistive technology arena, but if we take that database and give it to our friends at NIDRR and those that are working in the assistive technology arena, they can recognize those much better than they can today.

Mrs. MORELLA. That's great.

Would our industry representatives also want to put in the cost factor? I know some of these results are exceedingly costly.

Maybe when they are multi-manufactured, the costs will go down such as they did with computers.

But, you know, if that is any kind of an obstacle——

Mr. JACOBS. I'm glad I had a few seconds to think about it.

In answer to your question, I think that exclusivity of partnership is important. It's nice to believe that NCR and Hewlett Packard and IBM would partner with a federal lab and promote and further develop a product.

But we run our business on competitive advantage. I think one of the things that's very doable is arranging for exclusivity to the point where you would really engage the interest of big business.

There's a lot of economies of scale, I think, that could come into play there, latitude to negotiate a fair and equitable business agreement so that both the government and federal and private industry benefit financially from the arrangement.

Federal laboratories understanding private industry in the different markets and being able to segment the projects that are works-in-progress, or that may be sitting on the shelf due to lack of funding is also important.

I think that once you isolate the different groupings of technology that's being worked with, that one would need to promote the business benefits of collaborating with federal labs, which, in my opinion, is very, very easy to do.

I think this is an incredibly exciting opportunity. I don't really see any major barriers. There's probably some work to be done, but I don't see anything really standing in the way from a cost factor.

I think that the scenario of big business working together is nice, but in Steve Jacob's opinion, I don't think that's as easy to bring to bear as what it may seem on the surface.

I think we're in a very competitive marketplace, and anything that we could figure out that our competitors don't know about is just money in our back pocket.

So, I hope that answers your question.

Mrs. MORELLA. Thank you, great response. Thanks, Mr. Jacobs. Mr. Lahoud?

Mr. LAHOUD. I'm glad you raised the question of the price or the cost of these technologies. In our particular case, it's another good example of what is a problem, and that's what it costs people to get this technology from a small company like ours that has developed it to the point where it is today.

It's a very expensive product. The averaging selling price of our product for people with disabilities is $20,000.

As prohibitive as that sounds, there are over 120 people who are sources of funding that have enabled many devices to be placed.

I should have mentioned, and I think I mentioned in my written testimony, that one of the most critical reasons to develop the technology into the future is to reduce the price, is to find ways where this very valuable technology can be made much more affordable for the people that need to have it.

The other applications, of course, lead us into that. If we can find other uses for this technology, in many ways, the development of those uses will support the development of the products for people with disabilities.

As far as what could be done, I already mentioned, and at the risk of sounding repetitive, I guess I'd like to use an example, the example of drowsiness and fatigue prediction.

This is a major problem that's being addressed both publicly and in the private sector in this country. An enormous number of accidents, many more than half, probably 75 to 80 percent of accidents in this country are caused in one way or another by people being fatigued, not paying attention to their job.

So, if you look across the federal sector, you'll find the Department of Transportation, and within it there's the FAA and the Federal Highway Administration and the National Highway Traffic Safety Administration as the three major entities within DOT that are interested in fatigue prediction.

In the Department of Defense, the Army, the Navy, and the Air Force, all three, are extremely interested in predicting fatigue, and doing it in a high technology, unobtrusive way.

The Nuclear Regulatory Commission and the Department of Energy is interested because of problems that exist in nuclear power plant complex control rooms and so forth.

There are other government agencies, and if you were to somehow find a way, it's not like finding technologies that are duplicated. What are the resources that are being spent for a problem that is being solved?

I feel that if we can find all those different agencies that have within their budgets, X dollars to spend on predicting fatigue, we'll find that there's a lot of duplication of effort.

Any one agency will say that they can't afford to do everything that needs to be done, but yet if these resources could somehow be pooled——

The third answer to the question would have to do with—and you mentioned it very early in your remarks, Mrs. Morella—is the position that the United States has in high technology development relative to the rest of the world.

In our case of automatic eye tracking, it's very clear that the world has reached the stage where they recognize the value of this technology, the number of problems it can solve, the commercial potential associated with this technology.

The United States today definitely has a significant technological lead, but I see that lead beginning to dwindle from science in Japan and Germany, to name two countries.

I feel very strongly that we should become more aware of just what the international implications are for the United States maintaining some position of competition in this field of automatic eye tracking.

Mrs. MORELLA. Thank you. Mr. Hershberger?

Mr. HERSHBERGER. I think it would be good for us to consider this not necessarily so much as technology transfer as technology collaboration.

I think we all sort of envision or have the thought of finding some technology somewhere that instantly brings all kinds of benefits to people with disabilities.

Sometimes that is the case, but more often it's not. Often, even the cost of adapting the technology for people with disabilities is beyond the reach of the funds of the assistive device companies.

So I think what we often have is that those of us who work with assistive technology or work in the field of disabilities have a much greater knowledge of the needs that are faced by these people, but we don't have the funds to work on the problems, or we don't have the funds to work on the problems to the extent that the federal laboratories have.

So, I guess I would just see it as more of a collaborative effort between all those involved in trying to solve these problems.

Mrs. MORELLA. Certainly that would be desirable. I've often wondered whether or not having some kind of a cost/benefit ratio or a balance sheet where you point out how much you really save by making people's lives productive, and, you know, allowing them to fulfill a higher quality of life versus the expense of it—we tried to do that with mental illness, to point out, you know, this is really a health problem.

We can do much with enabling people to have a fuller life by virtue of investing in it. I think it's the same thing with the assistive technologies. If they can end up paying off exceedingly well—well, I'll get into that in a minute, but I wanted to recognize and offer any opportunity for questioning to our distinguished colleague, Mr. Ewing, who actually represents the University of Illinois.

Mr. EWING. Thank you very much, Madam Chairwoman.

I've really only been here for a short time, and I don't think I have a good, intelligent question for this very distinguished panel, but I'm learning by listening.

Thank you for the offer.

Mrs. MORELLA. Thank you, Mr. Ewing. I just wonder, what does the Federal Laboratory Consortium do to make sure that the word gets out, because that's something we hear a lot about, that we need better communication. You've mentioned that yourself, the need for awareness.

How do people find out? How do companies find out about the existing technology and the spinoffs? I wonder about what the FLC has done, can do, in terms of getting the word out?

Mr. BRAND. Well, I think in the past that we have done several publications. But as you know, the world and technology has changed very rapidly.

So, in the last 3 months we have been putting together a much-improved Website for the federal lab organizations, so that individuals in the private sector will be able to go on the Internet and locate federallabs.gov and locate the organizations that are members of the Federal Laboratory Consortium, and also locate their technologies through them.

We are certainly working with the other technology transfer organizations that are in this country to do the same thing in a similar fashion as part of that answer. Also, to my industry colleagues down here, it seem that the one barrier was the agency priorities.

When I addressed your answer earlier, saying that there were perhaps no legislative barriers, perhaps one of them we could do is to write some specific legislation language encouraging our exploration or pursuit of those technologies in the assistive technology transfer arena.

We have not done so, and that perhaps is a way to get the attention of the laboratories and scientists and engineers out there. From the communications standpoint, we have increased the awareness of assistive technologies.

Certainly in all of our publications, we plan to start a new technology transfer representative in each one of the laboratory's alert-o-grams. That's a coined-up phrase that I have coined up, but it's to alert our people in the laboratories of the assistive technology needs that are in the community. That will go to all 711 laboratories.

Those needs are also going to be on our Internet system that we use under AZINCOM. It's a communication system that goes to all of the laboratories.

If there is a particular need that we're looking for in the assistive technology transfer community, it will go to every one of those laboratories instantaneously.

Certainly, we need more awareness to the public and to the assistive technology transfer community; that's what I think. One of the arrangements we have recently formed is for the Consumer Assistive Technology Network.

It's a funded organization from NIDRR located in New Mexico. It's helping us locate technologies within the laboratories, and finding companies who need those particular laboratories and the awareness and success stories of those will be published very soon.

Mrs. MORELLA. You know, at the exhibition, Maryland was represented, the District of Columbia was represented, the national Department of Education, for instance, was represented.

Does every State have a vocational education, whatever it may be—have the assistive technology utilization, or a department that handles that?

Dr. Seelman?

Ms. SEELMAN. NIDRR administers the technology-related assistance for the Individuals With Disabilities Act which was passed by the Congress in 1988, recognizing the very problem that you're addressing now.

There are projects in the 50 States. However, there are some projects now sunsetting because there were limits, time limits which we're trying to address.

These projects have been extraordinarily successful in interfacing technology with the people who use it in systems change kinds of situations so that wheelchair companies, for example, can be put together with people who use them.

In every one of the States, there is an assistive technology project, just as there is in Maryland.

Mrs. MORELLA. Very good.

Did you have a comment on that? I might also ask you, because I was going to ask the industry people about the SBIR program, since Mr. Lahoud mentioned it, whether you find it helpful as industry people.

But do you have a response?

Mr. JACOBS. What I wanted to respond to was your original question concerning how best to move forward to make industry and rehabilitation engineering centers aware of the work being done in the federal labs.

In October, there is an opportunity coming up that we've been working with Scott Hall on at Wright Patterson Air Force Base, and with Mr. Brand. I forgot the co-chairman's name. Forgive me. Mr. Doug Blair.

When AT&T trivested—all the companies trivested, except one group called IDEAL, the Individuals With Disabilities Enabling Link, who are a group of about 500 engineers, that crossed Lucent, AT&T and MCI and were very proud to continue to be associated with each other in a capacity that doesn't compromise any intellectual property from any of the companies.

But in October we're having a joint conference in Atlanta that is going to be focused on technology, and how people with disabilities can enhance their careers, look forward to a much more positive lifetime than they may have in the past.

And we're going to be collaborating on the labs coming down and setting up a display that will have very high visibility to all three companies, engineers from all three companies.

I think that's a wonderful place to start, with three companies that each have their own market niche, three companies that are very big in size. I imagine there's going to be a lot of publicity about this event.

I think a lot of good things can come of this for everyone involved. So, that was going to be my comment, that any support that we could be provided will be very, very much appreciated.

Thank you.

Mrs. MORELLA. We'll try to do our part in getting the word out. That would be great to have that as a model.

Now, how about the SBIR Program? Just very briefly, Mr. Jacobs, have you had any contact with it at all?

Mr. JACOBS. Well, SBIR, I don't think, falls into our domain. We're about a $7 billion a year——

Mrs. MORELLA. You're too big.

[Laughter.]

Mr. JACOBS. But, personally, maybe I'll have to give them a call.

Mrs. MORELLA. I think you've had a success story with it; have you? This is a program we tend to look at in this Subcommittee.

Mr. LAHOUD. We've had several SBIR grants and contracts over the years, the more recent ones having to do with the automatic eye tracking. It's a wonderful program.

In the most recent years, it has put some additional effort into emphasizing multiple use technologies, and promoting technologies for people with disabilities. But in my opinion, more could be done to promote the development of technologies through SBIR that would assist people with disabilities.

It's a good resource, but I don't think they've gone far enough to emphasize how technologies should be also looked at from the standpoint of helping people with disabilities.

Mrs. MORELLA. You may want to get in writing to us, any ideas that you may have in that particular regard.

Mr. Hershberger?

Mr. HERSHBERGER. Prentke Romich Company has participated in a number of SBIR programs. We use it to aid what we're doing. It doesn't fund our entire research effort, but it does allow us, with specific programs, to get through the project much more quickly through additional funds.

I have had some experience with working with colleagues, particularly in the European community, using federal funds for assistive technology development.

I must say that the way that our system works, works very well in targeting the funds for projects that really do have the potential of having a benefit to people with disabilities.

Mrs. MORELLA. Dr. Seelman?

Ms. SEELMAN. I just wanted to restate the contrast between universal design and what we refer to as orphan technology, because of Steve's response. Orphan technology is usually a user population of about 250,000 or less.

Now, in the drug business, we have an Orphan Drug Act, which refers to that, and tax credits which helps the struggling companies.

Now, at the other end of the technology range is the Tech Act and the Tech Projects. They, too, must struggle with the orphan technology and who is going to pay for it.

The projects work with the schools. They work with the vocational rehabilitation agencies. They work with Medicaid Managed Care to try to support the costs on that end, in order to make sure that there is a revenue flow.

That's much of what the Tech Act does. it works in a systems change way with institutions within a State like Maryland. But I did want to reaffirm the two different orientations here; one being orphan technology and 250,000; and the other being universal design, more or less represented by Steve's presentation.

Mrs. MORELLA. I think you've all answered our questions quite well. What we've done is, the members of the Subcommittee will have an opportunity to pose any other questions to you in the hopes that you would respond, if that's okay.

I wanted to thank all of you for being here and for your presentations, and I look forward to also hearing from you. If there are any recommendations that occur to you subsequent to this hearing, please let us know so that we can make sure that we fulfill the awareness, communication, and do what we can to make sure that there is that collaboration.

I like that point, technology collaboration. I also wanted to thank the signers for doing such a wonderful job for such a long period of time.

So, now our Subcommittee hearing is now adjourned.

[Whereupon, at 3:40 p.m., the hearing was adjourned.]

○

APEX
TP_CF
9780160560460